75th Annual Meeting of the Northeastern Weed Science Society 2021

Proceedings – Volume 75

Online
4 – 7 January 2021

Editor:

Lane Heimer

ISBN: 978-1-7138-4212-5

Printed from e-media with permission by:

Curran Associates, Inc.
57 Morehouse Lane
Red Hook, NY 12571

Some format issues inherent in the e-media version may also appear in this print version.

Printed with permission by Curran Associates, Inc. (2022)

Northeastern Weed Science Society (NEWSS)
P.O. Box 25
Woodstown, NJ 08098
USA

Phone: 315-209-7580

northeasternweedscience@hotmail.com

Additional copies of this publication are available from:

Curran Associates, Inc.
57 Morehouse Lane
Red Hook, NY 12571 USA
Phone: 845-758-0400
Fax: 845-758-2633
Email: curran@proceedings.com
Web: www.proceedings.com

PROCEEDINGS - VOLUME 75

75th Annual Meeting of the

Northeastern Weed Science Society

January 4-7, 2021 – Virtual Meeting

Site of 2022 Joint Meeting*

January 3-6, 2022

Wyndham Gettysburg
Gettysburg, PA

Participating Societies:

Northeastern Weed Science Society (NEWSS)
Northeastern Regional Branch of
    American Society of Agronomy
    Crop Science Society of America
    Soil Science Society of America
American Society of Horticultural Science
    - Northern Division (NR-ASHS)

To order a copy of the Proceedings contact:
CURRAN ASSOCIATES, INC
57 Morehouse Lane, Red Hook, NY 12571
Ph: 1 845 758 0400 x112

§ Cover: The winning image in the 2021 NEWSS Photo Contest, submitted by Maggie Wasacz, Rutgers University, Asiatic Dayflower

Proceedings

of the

Seventy Fifth Annual Meeting

of the

Northeastern Weed Science Society

NORTHEASTERN WEED SCIENCE SOCIETY

75th 1946 - 2021 Anniversary

Lane Heimer, Editor

NORTHEASTERN WEED SCIENCE SOCIETY

2021 SUSTAINING MEMBERS

Platinum

BASF We create chemistry

syngenta

CORTEVA agriscience

Gold

Bayer CropScience

FMC

VALENT

Silver

UPL

Bronze

ACDS Research, Inc.

TeeJet Technologies

LABServices

For-Shore Weed Control

Table of Contents

NEWSS EXECUTIVE COMMITTEE – OFFICERS

President	**Caren Schmidt** BASF caren.schmidt@basf.com
President-elect	**Jacob Barney** Virginia Tech jnbarney@vt.edu
Vice President	**Wesley Everman** North Carolina State University wes_everman@ncsu.edu
Treasurer	**Larissa Smith** Syngenta Larissa.smith@syngenta.com
Past President	**Art Gover** Pennsylvania State University Aeg2@psu.edu

NEWSS EXECUTIVE COMMITTEE – MEMBERS

Membership Chair	**Thierry Besançon** Rutgers University thierry.besancon@rutgers.edu
Editor	**Lane Heimer** Maryland Department of Agriculture lane.heimer@maryland.gov
Public Relations	**John Kaminski** Pennsylvania State University kaminski@psu.edu
Research and Education Coordinator	**Matt Cutulle** Clemson University mcutull@clemson.edu
Sustaining Membership	**John Wallace** Pennsylvania State University Jmw309@psu.edu
CAST Representative	**Hilary Sandler** University of Massachusetts Cranberry Station hsandler@umass.edu
Graduate Student Representative	**Maria Gannett** Cornell University mag449@cornell.edu
Graduate Student Representative Elect	**Eric Jones** North Carolina State University Eajone22@ncsu.edu
WSSA Representative	**Rakesh Chandran** West Virginia University rschandran@mail.wvu.edu
Science Policy Director	**Lee Van Wychen** Weed Science Society of America lee.vanwychen@wssa.net

2021 NEWSS MEETING PROGRAM COMMITTEES

Program Committee	Chair:	Jacob Barney, Virginia Tech University
Agronomic Crops	Chair:	Vijay Singh, Virginia Tech University
Turfgrass, Ornamentals, and Non-Crop	Chair:	Matt Elmore, Rutgers University
Fruit, Vegetable, and Herb Crops	Chair:	Kurt Vollmer, University of Maryland
Research Posters	Chair:	Kate Venner, BASF
Student Oral Papers	Chair:	Dan Kunkel, AMVAC
Student Posters	Chair:	Anthony Witcher, Tennessee State University

2021 NEWSS COMMITTEES

Student Paper Contest	Chair:	Daniel Kunkel, Rutgers University
Student Poster Contest	Chair:	Anthony Witcher, Tennessee State University
Photo Judging Contest	Chair:	Hilary Sandler, University of Massachusetts
Site Selection Committee	Chair:	Wes Everman, North Carolina State University
Awards Committee	Chair:	Art Gover, Pennsylvania State University
Nomination Committee	Chair:	Lynn Sosnoski, Cornell University
Resolutions Committee	Chair:	Carroll Mosely, Syngenta
Collegiate Weed Contest	Chair:	Jacob Barney, Virginia Tech
Audit Committee	Chair:	Larissa Smith, Syngenta
Archives Committee	Chair:	Art Gover, Pennsylvania State University
Past Presidents Committee	Chair:	Art Gover, Pennsylvania State University
Endowment Committee	Chair:	Art Gover, Pennsylvania State University

2021 NEWSS AWARDS

NEWSS Fellow Award – Dr. Gregory Armel

Award of Merit – Quintin Johnson

Outstanding Researcher Award –

Dr. Sandeep Rana

Outstanding Educator Award –

Dr. Thierry Besançon

Outstanding Educator Award -

Dr. John Kaminski

M. Garry Schnappinger Service Recognition Award – Dr. Kurt Vollmer

Robert D. Sweet Outstanding Ph.D. Student Award – Dr. Theresa Reinhardt Piskáčková

Robert D. Sweet Outstanding M.S. Student Award – Eric B. Scruggs

2021 NEWSS MEETING CONTEST WINNERS

STUDENT PAPER

1st place paper – Maria Gannett, Cornell University, "Soil Carbon Amendments as a Microbially Mediated Weed Management Tool"

2nd place paper – Vasiliy Lakoba, Virginia Tech University, "Home Climate and Land Use Affect Seed, but not Rhizome Cold Tolerance of a Perennial Invader"

2nd place paper – Matthew Spoth, Virginia Tech University, "Soybean Performance After Chaff Lining at Wheat Harvest"

Honorable Mention – Maggie Wasacz, Rutgers University, "Crop Response to Simulated Dicamba Drift in Field -Grown Snap Beans"

Honorable Mention – Uriel Menalled, Cornell University, "Organic No-Till Soybean Seeding Rate and Nitrogen Fertilization Effects on Weed Suppression."

Honorable Mention – Levi Moore, North Carolina State, "Evaluating Electrical and Mechanical Methods for Palmer Amaranth (Amaranthus palmeri) Control"

NEWSS STUDENT LIGHTNING TALK CONTEST

1st place: Marshall Warren, North Carolina State University, "Evaluation of Chopped Miscanthus Biomass as a Mulch for Weed Control in Container Grown Nursery Crops"

2nd place: Maggie Wasacz, Rutgers University, "Imaging Analysis to Quantify Leaf Deformation in Response to Sublethal Rates of Dicamba"

3rd place: Sandra Ramsey, North Carolina State University, "Characterization of Herbicide Safety in *Brassica carinata* (A,) Braun in the Piedmont and Coastal Plains of North Carolina"

PHOTO CONTEST

1st place photo – Maggie Wasacz, Rutgers University, Asiatic Dayflower

2nd place photo – Dan Kunkel, AMVAC, Poison Ivy at the Jersey Shore (NJ)

3rd place photo – Sandeep Rana, Bayer Crop Science, Velvetleaf Seedpod at Bayer's Galena, MD Research Station

UPDATE ON 2020 WEED SCIENCE RESEARCH IN THE IR-4 ENVIRONMENTAL HORTICULTURE PROGRAM. CL Palmer*, E Vea; IR-4 - Rutgers University, Ne Brunswick, NJ, IR-4 - Rutgers University, Crownsville, MD

ABSTRACT

In 2020, the IR-4 Environmental Horticulture Research Program sponsored research on several weed science projects: pre- and postemergent herbicide crop safety with over-the-top in-season applications, efficacy of fatty-acid herbicides, management of Nostoc, and management of liverwort. For the over-the-top herbicide crop safety screening the goal was to develop lists of crops where applications would not harm woody and herbaceous perennials grown in container nurseries. Products tested included Basagran (bentazon), Dimension 2EW (dithiopyr), Fiesta Herbicide (FeHDTA), Fortress (isoxaben + dithiopyr), Freehand G (pendimethalin + dimethenamid-p), Fuerte (OHP1702V), Gallery SC (isoxaben), Gemini Granular (prodiamine + isoxaben), Marengo 74SC (indazaflam), Pendulum G (pendimethalin), Ronstar G (oxadiazon), SP1770, and Tower EC (dimethenamid-p). Applications were made at dormancy (preemergents) or when leaves were fully expanded (postemergents) and then approximately 6 weeks later for all products. Basagran was screened on 20, Dimension EW on 4; Fiesta on 5; Fortress on 8; Freehand on 1; Fuerte on 13; Gallery on 8; Gemini Granular on 5; Marengo on 27; Pendulum G on 10; Ronstar G on 2; and Tower was screened on 13 crops. The results from this research will aid in the development of product labels and will help growers and landscape care professionals make more informed product choices.

WEED SEED GERMINATION FOLLOWING UP TO 9 MONTHS OF SUBMERSION IN IRRIGATION PONDS. AR Shiffer*, JC Neal, A LeBude, J Altland, C Harlow; North Carolina State University, Raleigh, NC, NC State University, Mills River, NC, USDA-ARS, Wooster, OH

ABSTRACT

Nursery crops producers use open ponds as source water for irrigating container grown plants. After irrigation and rain events, production tailwater is captured in these ponds and often reused for irrigation without filtration. Runoff directed to the pond can become contaminated with weed seeds from various sources. Ponds have large surface areas, so it may take days or weeks before weed seeds can be reintroduced into production through irrigation. Whether or not weed seeds are still viable after being introduced into the pond has not been studied widely in nursery production. Therefore, this study investigated the submersion of four common weed species in pond water at four nurseries in eastern North Carolina, U.S. Seeds were submerged for 7, 14, 21, 30, 60, 90, 120, 180, and 360 days beginning 20 Nov 2019, then retrieved and germinated for 30 days. Seeds of all species showed germination throughout the study, notably *Cardamine flexuosa* maintained 70% germination up to at least 360 of submergence.

CONVENTIONAL AND ORGANIC ALTERNATIVES TO GLYPHOSATE. J Derr*; Virginia Tech, Virginia Beach, VA

ABSTRACT

Glyphosate has been a standard for site preparation, as well as nonselective control of annual and perennial in established ornamental plantings and many other areas. It has been suggested that glyphosate may cause non-Hodgkin lymphoma or other health issues. This has resulted in an interest in alternatives to glyphosate, including organic options. The objectives of this research were to evaluate alternative chemicals to glyphosate, including conventional and organic options and to determine speed of action as well as long term control. Treatments that were evaluated in a container and a field trial were: a Nontreated, Roundup ProMax (glyphosate), Cheetah Pro (glufosinate), Diquat plus Capsil surfactant, Natural Armor RTU, Weed Zap, Weed Slayer plus Agro Gold, Avenger Weed Killer, Scythe (pelargonic acid), and Weed Pharm (acetic acid) applied undiluted. In the container trial, glyphosate and glufosinate gave excellent control of all six weed species at 19 DAT. Weed Slayer plus Agro Gold provided excellent control of longstalked phyllanthus (Phyllanthus tenellus Roxb.), good control of southern crabgrass [*Digitaria ciliaris* (Retz.) Koel], Virginia buttonweed (Diodia virginiana L.) and compressed sedge (Cyperus compressus L.), and fair control of bermudagrass [Cynodon dactylon (L.) Pers.] and yellow nutsedge (Cyperus esculentus L.). Diquat gave excellent control of longstalk phyllanthus, yellow nutsedge, and compressed sedge but poor control/regrowth of the other weed species. Pelargonic acid gave fair control of longstalk phyllanthus but poor control of the other weeds. The other treatments caused injury soon after application, but the weed species were able to outgrow the damage. Repeat applications would be need for these treatments. In the field trial, glyphosate, glufosinate, and Weed Slayer + Agro Gold gave complete control of southern crabgrass at 17 and 28 DAT. The other treatments caused significant burning of southern crabgrass foliage at 4 DAT but this weed outgrew that injury by 17 DAT. Glyphosate and glufosinate gave excellent control of yellow nutsedge at 17 and 28 DAT, with good control seen with diquat and fair control with Weed Slayer + Agro Gold. The other treatments did not provide acceptable control of yellow nutsedge. Glufosinate is an effective conventional alternative to glyphosate while Weed Slayer + Agro Gold appears to be an acceptable organic alternative. Of the conventional herbicides, glufosinate gave comparable weed control to glyphosate while diquat and pelargonic acid gave lower overall weed control. Of the organic treatments, Weed Slayer + Agro-Gold gave good to excellent weed control, depending on weed species. The other organic treatments provided contact injury but weeds were able to outgrow the damage.

FACTORS AFFECTING THE EFFICACY OF NONSELECTIVE SHORT-CHAIN FATTY ACID HERBICIDES. AF Senesac*; Cornell Cooperative Extension of Suffolk County, Riverhead, NY

ABSTRACT

A field trial was established in 2020 under the trellis area of an established vineyard at LI Horticultural Research & Extension Center. The resident vegetation consisted of small and large crabgrass (Digitaria spp.), white clover (Trifolium repens) and a strip of established fescue (Festuca spp.). Treatments consisted of comparisons of Homeplate 80% a.i. (caprylic acid 44% + capric acid 36%) at 6% and 9% solutions and applied at 50 and 100 GPA. The 6%-50 GPA treatment was evaluated with and without 1% and 5% Biolink Acidifier (citric acid). AXXE (ammonium nonanoate) was also evaluated at 9% solution applied at 50 GPA. During treatment preparation, the Homeplate treatments were pulse blended for 10 seconds using a bench blender. This provided sufficient blending of the active ingredients with the water carrier to provide a homogenous application of the treatment. The plots were visually evaluated for injury (weed control) at 1, 3, 7, 14, 21, and 28 days after treatment (DAT). The results indicate that there were numerical but few significant differences between the two rates of Homeplate when applied in either 50 or 100 GPA. These four treatments provided good to excellent crabgrass injury (control) for the first 14 days after treatment. After 14 DAT, the crabgrass began to recover partially but some level of control was maintained until the end of the trial. Injury to white clover and fine fescue followed a similar pattern of high level of initial injury followed within two to four weeks by partial recovery. An interesting result was observed with treatments where acidifier was added to Homeplate. An anecdotal observation had suggested that adding acidifier would improve the performance of Homeplate. However, the acidifier actually appeared to decrease the performance of Homeplate. At each evaluation, the acidified Homeplate treatments provided significant lower levels of injury to the crabgrass, clover, and fine fescue. A comparison of the Homeplate at 9% solution- 50 GPA to AXXE at 9% solution- 50 GPA indicated that Homeplate performed significantly better for the first 14 days after treatment for all three test species. Based on these results, it appears that for Homeplate, the lower rate and gallonage that was tested provides statistically the same level of initial control as the higher ones in the first two weeks after treatment.

EFFECT OF MULCHES AND PRE-EMERGENCE HERBICIDES ON GROWTH OF ROOTED CUTTINGS AFTER TRANSPLANT. AL Witcher, I Poudel*; Tennessee State University, McMinnville, TN

ABSTRACT

Weeds are a major problem in the cutting propagation environment, competing for resources and reducing crop growth and quality. Hand weeding is the most common method for controlling weeds in nursery crop propagation, but it is time-consuming and costly due to high labor cost. The objective was to determine the effect of mulches (rice hulls and pine pellets) and pre-emergence herbicides (isoxaben, oxyfluorfen + oxadiazon, isoxaben + dithiopyr, and indaziflam) on rooting of cuttings and growth after transplant. Terminal and sub-terminal 9-10 cm stem cuttings of butterfly bush (*Buddleia davidii* 'Nanho Blue') was collected from container grown stock plants. Cuttings were then stuck in 6.35 cm containers treated with pre-emergence herbicides and mulches then were kept under intermittent mist. After sufficient rooting had occurred (based on the non-treated control), 8 cuttings (8 replications per treatment) were randomly selected and transplanted to trade gallon (2.4 L) containers filled with a pine bark substrate, then placed on a greenhouse table, and irrigated daily. Shoot growth (height and width), leaf greenness (SPAD), and substrate pH and electrical conductivity (EC) were recorded monthly for 2 months. At termination of the study, all plants were harvested to record shoot and root dry weight. The remaining 17 cuttings (from each treatment per species) were harvested and data were collected including rooting percentage and shoot and root dry weight. All data were analyzed with linear models using the GLIMMIX procedure of SAS and differences between treatment means were determined using the Shaffer Simulated method ($P < 0.05$). Rooting percentage was (over 80%) for all the treatments except isoxaben (20%) and isoxaben + dithiopyr (70%). Isoxaben also resulted in lower root and shoot growth of butterfly bush cuttings. The remaining treatments did not have any negative effect on root and shoot growth of butterfly bush cuttings. After transplant, all of the treatments (indaziflam, oxyfluorfen + oxadiazon, rice hulls and pine pellets) were similar to the non-treated control for butterfly bush shoot growth, root and shoot dry weight, SPAD, number of flowers, and substrate EC and pH. Isoxaben and isoxaben+dithiopyr are not currently labeled for use on butterfly bush and isoxaben has been reported to cause injury to butterfly bush in container production. Nevertheless, the remaining treatments did not cause any negative effects on butterfly bush root and shoot development. In conclusion, certain pre-emergence herbicides and mulches may be viable alternatives for weed control in propagation, but additional crop species should be evaluated.

WEED AND TURFGRASS RESPONSE TO DURATION OF DIRECT FLAME. S Askew*, CG Goncalves; Virginia Tech, Blacksburg, VA

ABSTRACT

Only one preliminary report of a non-repeated study could be found regarding flame weed control in ornamental turf ('Patriot' hybrid bermudagrass). Studies conducted in rangeland and prairie ecosystems suggest that evolved selectivity to fire could be a viable weed control option for turfgrass systems. From a meta-analysis of 500 search returns via google and google scholar, we found that flaming had a high likelihood of occurring in consumer searches for organic weed control options in ornamental turf. Based on the paucity of scientific evidence to support Internet trends for flame weed control topics in ornamental turfgrass, we conducted studies to assess the response of different cool-season turfgrass species and weeds to duration of flame exposure. Field trials were conducted during the 2020 growing season at Virginia Tech's Glade Road Research Facility in Blacksburg, VA, to examine the efficacy of duration of flame exposure for weed control in tall fescue (*Festuca arundinacea*), Kentucky bluegrass (*Poa pratensis*), perennial ryegrass (*Lolium perenne*), and hard fescue (*Festuca* spp.). Treatments included uniform flaming of 6 ft^2 turf plots for 1, 2, 3, and 4 minutes at approximate speeds of 0.4, 0.2, 0.1, and 0.05 mph, respectively using a propane torch with a 4-inch diameter head. A nontreated check was included for comparison. Data on weed control and turfgrass recovery were collected and analyzed. Almost all vegetation was incinerated by flame exposure regardless of exposure level. Tall fescue and Kentucky bluegrass recovered from flame treatments quickly with acceptable turf cover reestablished in under four weeks. However, perennial ryegrass and hard fescue recovery were flame-duration dependent and generally unacceptable. Corn speedwell and common chickweed were completely controlled regardless of flame duration. Annual bluegrass and white clover were controlled up to 85% depending on flame duration. Common dandelion, buckhorn plantain, horseweed, and broadleaf dock were initially burned back by flaming, but recovered by 49 DAT regardless of flame duration.

ANTAGONISTIC MIXTURES OF HERBICIDES AND FUNGICIDES IN TURFGRASS. JM Peppers*, S Askew; Virginia Tech, Blacksburg, VA

ABSTRACT

Herbicide efficacy can be affected by the addition of additional agrochemicals in a tank mixture. Previous research indicates that the addition of fungicides to herbicides can reduce herbicidal injury to turfgrass. However, the effect that these combinations have on weed control has not been examined in turfgrass. Research was conducted in Blacksburg, Virginia to evaluate smooth crabgrass (*Digitaria ischaemum* Schreb.) control with five herbicides alone and in combination with five fungicides. The five herbicides examined in this study were fenoxaprop, pinoxaden, quinclorac, siduron, and topramezone applied at 17.5, 70.6, 578, 6725, and 6.13 g ai ha^{-1}, respectively. These herbicides were applied alone and in combination with chlorothalonil, propiconazole, azoxystrobin, iprodione, and fluazinam at 7353, 501, 609, 3060, and 803 g ai ha^{-1}, respectively. These treatments were applied to a monostand of smooth crabgrass that was between 3 and 7 tillers at the time of the initial application and these treatments were reapplied at 14 days after the initial treatment (DAIT). At 42 DAIT, fenoxaprop alone controlled smooth crabgrass 64%, but combinations of fenoxaprop with any of the fungicides reduced smooth crabgrass control to <10%. Smooth crabgrass control with pinoxaden was significantly affected by all five fungicides as well. At 42 DAIT, pinoxaden alone controlled smooth crabgrass 74%, while pinoxaden plus fluazinam controlled smooth crabgrass 39%. The other four fungicides reduced pinoxaden control of smooth crabgrass to <20%. The addition of fungicides did not significantly affect the activity of quinclorac and siduron. The addition of chlorothalonil and fluazinam reduced smooth crabgrass control with topramezone from 96% to 66 and 71%, respectively compared to topramezone alone. In this study, each fungicide reduced the smooth crabgrass control of at least two herbicides. However, quinclorac and siduron were unaffected by the addition of fungicides in a tank mixture. The two graminicides examined were the most sensitive to the addition of fungicides in this study. Future research should further examine the effects of fungicide tank mixtures with herbicides on other weed species.

NON-CHEMICAL APPROACHES TO ANNUAL BLUEGRASS CONTROL IN PERENNIAL RYEGRASS AND TALL FESCUE. CG Goncalves*, JM Peppers, JM Craft, J Brewer, S Askew; Virginia Tech, Blacksburg, VA

ABSTRACT

Organic weed control options for turfgrass systems remain in high demand. For annual bluegrass (*Poa annua*), the problem has intensified because of resistance to many once-effective synthetic herbicides. Although there are few products and strategies that are labeled as organic options, the viability, practicality, and efficiency of these alternative strategies for weed control in turfgrass system is poorly understood. Field trials were conducted during the 2020 growing season at Virginia Tech's Glade Road Research Facility in Blacksburg, VA, to examine the efficacy of alternative strategies for annual bluegrass control in perennial ryegrass and tall fescue turf. The organic treatments included hand weeding, boiling water, flaming, sod replacement, solarization (4000 W/m^2), solarization (8000 W/m^2), and shading (six weeks). Also, were included a selective herbicide (Methiozolin) and a non-selective herbicide (Glufosinate) as a standard treatment, and a nontreated check. All treatments were replicated four times and herbicides were applied with a CO_2-powered boom sprayer calibrated to deliver 40 gal/ac. Results from these trials indicate that manual weed control and sod replacement may be great options, but may be cost prohibitive. Shading was minimally effective at controlling annual bluegrass and may require more than six weeks disruption of turf use. Solarization appears to be seasonally dependent such that cool, low-radiation, conditions in spring will be less effective than warm, highly radiant summer conditions. Flaming and Boiling-water treatments controlled all turf vegetation greater than or equal to glufosinate. Among these nonselective treatments, boiling water resulted in less subsequent weed germination than flame and overall quicker tall fescue and perennial ryegrass recovery compared to flame or glufosinate.

TRICLOPYR MIXTURES FOR IMPROVED CREEPING BENTGRASS RESPONSE TO TOPRAMAZONE. D Koo*, J Brewer, S Askew; Virginia Tech, Blacksburg, VA

ABSTRACT

Topramezone (Pylex) is a hydroxyphenylpyruvate dioxygenase (HPPD) inhibitors registered on golf courses, sod farms, and residential turfgrass. Topramezone controls several broadleaf and grass weeds including crabgrass and goosegrass in post-emergence. It is registered in several cool- and warm-season turfgrasses including creeping bentgrass. The potential common injury on creeping bentgrass by topramezone is bleaching symptoms. Bleaching symptoms do not damage the health of the turf, however, it can arouse aesthetic concerns for professional turfgrass managers. In these studies, triclopyr admixture with topramezone is tested on the creeping bentgrass to evaluate the bleaching symptoms reduction. Field trials were conducted from August to December 2020 at Virginia Tech's Glade Road Research Facility in Blacksburg, VA. The recommended rate of topramezone on creeping bentgrass (0.25 fl oz/A) and double rate (0.5 fl oz/A) were applied with and without an admixture of 1 fl oz/A of triclopyr (Turflon Ultra Ester). Two applications with three weeks intervals were made to refer to the typical program for controlling mature crabgrass and goosegrass. All treatments were replicated four times and herbicides were applied with a CO_2-powered boom sprayer calibrated to deliver 30 gal/ac. Visual evaluation of creeping bentgrass injury and bleaching was conducted. Other field trials were conducted in August 2020 at three different golf courses in Virginia. The recommended rate of topramezone on creeping bentgrass (0.25 fl oz/A) and double rate (0.5 fl oz/A) were applied with and without an admixture of 0.75 fl oz/A of triclopyr (Turflon Ultra Ester). Other triclopyr containing herbicides (4 Speed XT, Cool Power) which is registered in creeping bentgrass for the golf course were also tested as an admixture in equivalent to the rate of the single triclopyr product. All treatments were replicated four times and herbicides were applied with a CO_2-powered boom sprayer calibrated to deliver 40 gal/ac. Visual evaluation of creeping bentgrass injury and bleaching was conducted. The result indicates the bleaching on creeping bentgrass caused by topramezone 0.25 fl oz/A can be reduced by adding triclopyr 1 fl oz/A in the sequential application. Different triclopyr containing herbicide registered in creeping bentgrass for the golf course shows similar injury and bleaching reduction to single triclopyr containing product.

HERBICIDE AND TALL FESCUE INTERSEEDING PROGRAMS FOR FALSE-GREEN KYLLINGA (KYLLINGA GRACILLIMA) CONTROL IN TURFGRASS. MT Elmore*, DP Tuck; Rutgers University, New Brunswick, NJ

ABSTRACT

A field experiment was conducted from 2019 to 2020 to evaluate late summer applications of various herbicides in combination with tall fescue (*Festuca arundinacea*) seeding for false-green kyllinga (*Kyllinga gracillima*) control. The site consisted of a mature stand of false-green kyllinga (>95% cover in mid-summer) that had been managed in 2018 and 2019 to nearly eliminate all other perennial species. Treatments consisted of seven herbicide treatments and a non-treated control in a complete factorial with tall fescue overseeding or no seeding. Herbicide treatments consisted of halosulfuron-methyl (70 g ha^{-1}), imazosulfuron (420 g ha^{-1}), pyrimisulfan (70 g ha^{-1}), and sulfentrazone + carfentrazone (280 + 31 g ha^{-1}) applied singly 4 weeks before tall fescue seeding and on the day of seeding. Glyphosate (220 g ae ha^{-1}) applied on the day of seeding was also included. Herbicide treatments were applied on 19 August 2019, 4 weeks before seeding (WBS), and 16 September 2019 (day of seeding) to 1.2 by 3.0 m plots using a CO_2-powered single nozzle boom with 410 L ha^{-1} of water carrier though an AI9504EVS nozzle. Imazosulfuron and halosulfuron-methyl were applied with NIS at 0.25% v/v. Tall fescue ('Winning Colors' blend) was seeded at 360 kg PLS ha^{-1} on 16 September 2019 using a shaker jar to a 1.5 m strip in each replicate. Preceding seeding, strips were core cultivated with 1.5 cm diameter hollow tines on a 3.8 cm spacing and vertically mowed to cut 0.6 cm slits on 2.5 cm spacing. Starter fertilizer (12 N - 24 P_2O_5 - 12 K_2O) was applied at 50 kg N ha^{-1} one week after seeding. Plots were arranged in a strip-plot randomized complete block design with three replications. False-green kyllinga control was evaluated visually on a 0 (i.e., no control) to 100 (i.e., complete control) percent scale relative to a non-treated control in June, July, and August 2020. A grid intersect count was conducted in July 2020. Turfgrass quality was evaluated in seeded plots on a 1 (poor) to 9 (best) scale. Tall fescue seedling injury was evaluated 1 month after seeding on a 0 (no injury) to 100 (complete death) scale. Data were analyzed using the GLIMMIX procedure in SAS (P = 0.05) and Fisher's protected LSD test was used to separate means. A herbicide-by-seeding interaction was detected for false-green kyllinga control on all three rating dates. In the absence of herbicide, tall fescue seeding provided 80% control in June, but 23% in August. Halosulfuron and imazosulfuron applied 4 WBS and glyphosate provided similar control (= 93%) in June regardless of tall fescue seeding treatment. By August these treatments provided = 50% control without tall fescue seeding, but >90% control with seeding. On the same date, sulfentrazone + carfentrazone applied 4 WBS provided 43% control with seeding, less than halosulfuron and imazosulfuron but more than the non-herbicide control with seeding (23%). Imazosulfuron, halosulfuron, and sulfentrazone + carfentrazone caused >60% tall fescue seedling injury when applied on the day of seeding which reduced turfgrass quality in May 2020. Interestingly, imazosulfuron applied 4 WBS resulted in 43% tall fescue seedling injury and poorer turfgrass quality (6.7) compared to the non-treated (7.8) in May 2020. By June, only imazosulfuron and sulfentrazone + carfentrazone applied at seeding reduced turfgrass quality. This research demonstrates that tall fescue seeding in conjunction with herbicide application improves false-green kyllinga control compared to either practice alone.

EFFICACY OF POST-EMERGENCE HERBICIDES AS ALTERNATIVES FOR GLYPHOSATE.

AL Witcher*[1], C Marble[2]; [1]Tennessee State University, McMinnville, TN, [2]University of Florida, Apopka, FL

ABSTRACT

Glyphosate is an important weed control tool for landscape professionals and in nursery crop field production due to low cost, broad spectrum weed control, and low environmental impact. Concerns over glyphosate resistant weeds and reported health issues attributed to glyphosate have led to increased interest in glyphosate alternatives. The objective of our work was to evaluate alternative post-emergence herbicides for efficacy of *Oxalis stricta* at two growth stages. Containers (14 cm diameter) were filled with substrate (4 pine bark : 1 peatmoss), placed on an outdoor container pad, and irrigated daily. Thirty weed seeds were sown in each container (8 containers per treatment) and containers were arranged in a randomized complete block design. At 6 and 10 weeks after sowing (WAS), herbicides were applied (high labeled application rate; application volume varied by herbicide) to separate sets of plants (6 and 10 WAS) using a handheld CO_2 sprayer. Synthetic herbicides (glyphosate, glufosinate, and diquat) and non-synthetic herbicides (pelargonic acid, caprylic+capric acid, ammonium nonanoate, acetic acid - 20%, eugenol, and d-limonene) were used. Phytotoxicity ratings (0-100% compared to a non-treated control) were assessed at 1, 2, and 4 weeks after treatment (WAT) and shoot dry weight was collected at 5 WAT. Many of the herbicides exhibited significant phytotoxicity at 1 WAT, yet plants treated with some non-synthetic herbicides tended to re-grow resulting in reduced control by 4 WAT. For the 6 week old plants, glyphosate, glufosinate, diquat, and eugenol provided 90% or greater phytotoxicity at 4 WAT. For the 10 week old plants, 90% or greater phytotoxicity was observed at 4 WAT for glyphosate, glufosinate, diquat, and pelargonic acid. All herbicides tended to provide increased control on the older (10 WAS) plants, likely due to the lack of vigor in the later growth stage. Shoot biomass was reduced by at least 73% for glufosinate and diquat for the 6 and 10 WAS plants, while lower biomass compared to the non-treated control was also observed for pelargonic acid and acetic acid for the 10 WAS plants. Future research with non-synthetic herbicides should focus on smaller weeds to reduce potential for regrowth after application. Overall, the synthetic herbicides glufosinate and diquat along with the non-synthetic herbicide pelargonic acid are acceptable alternatives to glyphosate for controlling *Oxalis stricta.*

AN ANALYSIS OF ALL US STATE NOXIOUS WEED LISTS: A PATCHWORK OF APPROACHES AND LOTS OF REGULATED NATIVE SPECIES. J Barney*, VT Lakoba, R Brooks, DC Haak; Virginia Tech, Blacksburg, VA

ABSTRACT

The number of weedy and invasive plants continues to increase globally. Once these plants establish, they become nearly impossible to eradicate and expensive to manage. Thus, prevention is the most effective tool for mitigation, which is instituted by both state and federal governments through noxious weed lists. Despite their near ubiquity, there has not been a comprehensive analysis of these regulated species throughout the US in >2 decades. Throughout the US, 46 states maintain noxious weed lists with a total of 1249 unique species, of which 48% are non-native, 40% are native, and 12% are not yet introduced to the US. Most of the listed species are already present in the listing state, precluding their ability to prevent new introductions. Also, a large number of entire genera, including some crop genera (eg, Brassica) are listed, as well as many native species - all of which complicates the utility and application of these regulations.

PRE-PLANT WEED SUPPRESSION CAN PROVIDE RESIDUAL ESTABLISHMENT BENEFITS FOR PLANTED SEEDLING TREES. AE Gover*, EN Weaver; Penn State, University Park, PA

ABSTRACT

Reforestation or afforestation to create riparian forest buffers is a challenging and costly endeavor, and we constantly seek readily adoptable operations to improve establishment success. Two such experiments were conducted beginning 2012 and 2015. The 2015 experiment examined the effect of pre- and post-plant weed suppression on establishment of tulip poplar (*Liriodendron tulipifera)* planted as 1-0 containerized seedlings and maintained in 1.5 m, rigid polypropylene shelters (tree tubes). The two experimental sites were properties of the Natural Lands Trust, the Crows Nest Preserve, Elverson, PA; and Stroud Preserve, West Chester, PA. Treatments included pre-plant treatment (yes/no) and post-plant treatments for 0, 2, or 4 years after planting, installed as a completely random design with a factorial treatment arrangement, each treatment replicated eight times, with individual trees as the experimental unit. Pre-plant treatments of glyphosate plus pendimethalin plus sulfometuron at 2.2 plus 4.4 plus 0.023 kg/ha, respectively, were applied October 29, 2015, to 1.8 m by 1.8 m planting spots. Crows Nest had been a hay field and was primarily perennial cool-season grasses with a mix of forb species and scattered exotic shrubs, while Stroud was maintained as a utility turf, dominated by perennial cool-season grasses. Seedlings were planted April 26 and May 6, 2016, at Crows Nest and Stroud, respectively, and had an average height of 47 cm at planting. Post-plant treatments were applied immediately after planting, and each spring according to assigned treatment, using the same mixture as the pre-plant application. At planting, and at the conclusion of each growing season, tree height was measured at the highest viable bud, and tree diameter was measured at 15 cm 2016 through 2018, and as DBH (1.4 m) in 2019. These metrics were used to calculate stem volume in cm^3 ($diameter^2$*height). Stem volumes from both sites were analyzed using a mixed model with site as a random variable. For 2016 and 2017, the data for 2- and 4-yr post-plant treatment stems were combined as a single post-plant treatment, and the full 0-, 2-, and 4-yr set of post-plant treatments were analyzed for 2018 and 2019. Annual mortality and cumulative survival were analyzed each season and was never a treatment effect, so dead stems were treated as missing values. Average survival as the end of the 2019 growing season was 64 percent. There was no significant interaction between pre- and post-plant treatments for stem volume for any growing season. Post-plant treatment was not a significant effect any growing season, though tree response followed the expected trend of more growth with more weed suppression. At the conclusion of 2019, average stem volume for trees treated for 0-, 2-, or 4-years after planting was 1255, 1432, and 1557 cm^3, respectively. Pre-plant treatment effect was significant 2016-2018 (p-values 0.0001, 0.0003, and 0.0033, respectively), and had a p-value of 0.11 for 2019, when stems receiving pre-plant weed suppression had an average volume of 1764 cm^3, and untreated stems averaged 1061 cm^3.

CHARACTERIZATION OF HERBICIDE SAFETY IN BRASSICA CARINATA (A.) BRAUN IN THE PIEDMONT AND COASTAL PLAINS OF NORTH CAROLINA. SA Ramsey*, AR Post, T Reinhardt Piskackova, ME Camacho, RG Leon; North Carolina State University, Raleigh, NC, Czech University of Life Sciences Prague, Prague 6 - Suchdol, Czechia (Czech Republic)

ABSTRACT

Brassica carinata (A.) Braun, or carinata, is an oilseed crop that is currently being developed for biofuel production. Southeastern growers have been interested in growing carinata as a winter crop because of biofuel industry demand and potential use as a rotational crop. As a new crop, there are no herbicides registered for use in carinata, so a preliminary screening was used to identify herbicides for safe use in carinata. Thus, the main objective of this study was to assess the safety of select preemergence and postemergence herbicides at varying rates on carinata seedling establishment and plant growth. The preemergence herbicides used were diuron, napropamide, and clomazone. The postemergence herbicides used were simazine and clopyralid and were applied at the 3- to 5-leaf stage. The rates that were tested were 0.25X, 0.5X, 1X, 2X, 4X and 8X the recommended label rate for each herbicide. The results indicated that despite high levels of bleaching and chlorosis, clomazone did not reduce yield at the 1X and lower rates compared with the non-treated control. Conversely, diuron caused almost complete carinata mortality and yield loss at most rates, so this herbicide could be used for control of volunteer carinata in rotational crops. Napropamide and clopyralid seem to be safe at label rates, but more research is needed due to high levels of variation depending on location. In the future, further investigation of the development of cultural and mechanical approaches are needed to provide growers with more weed control options for carinata production.

SEED TREATMENT SAFENS PRE-EMERGENT HERBICIDES IN BRASSICAS. GA Caputo*, MA Cutulle; Clemson University, Clemson, SC, Clemson University, Charleston, SC

ABSTRACT

The use of safeners applied as seed treatments to reduce injury from soil applied herbicides could improve weed management in brassica crops. An experiment was conducted to evaluate the dose-response of turnips and collards to *S*-metolachlor, pyroxasulfon, halosulfuron, and mesotrione when crop seeds were treated using melatonin, 24-Epibrassinolide, and ascorbic acid (AsA). Our results had demonstrated a low tolerance to herbicide by both crops, inhibiting germination in most all doses of pyroxasulfon, halosulfuron and, mesotrione, without seed treatment. When treated with melatonin, 24-Epibrassinolide, or AsA, herbicide tolerance improved, permitting germination on soils treated with pyroxasulfon, halosulfuron, and mesotrione. However, herbicide injury was superior to acceptable levels. Plants exhibit greater tolerance to *S*-metolachlor compared to other herbicides. Seed treatment significantly reduced the damage caused by this herbicide for both crops. These results suggest that the use of seed treatment using plant hormones and AsA could improve turnip and collard tolerance to PRE applications.

EFFICACY OF DICAMBA AND GLUFOSINATE APPLICATIONS ON PERVASIVE ROW CROP WEEDS OF NORTH CAROLINA. EA Jones*, DJ Contreras, M Fajardo Menjivar, RG Leon, W Everman; North Carolina State University, Raleigh, NC

ABSTRACT

Experiments were conducted to determine the efficacy of dicamba and glufosinate applications on two pervasive weeds of North Carolina, carpetweed (*Mollugo Verticillata*) and large crabgrass (*Digitaria sanguinalis*). Herbicide treatments (non-treated, dicamba, glufosinate, and dicamba+glufosinate) were applied the weeds at two different sizes (carpetweed: 2 and 5 cm, large crabgrass: 10 and 20 cm) at two locations. All herbicides were applied in a factorial arrangement with the inclusion of sequential applications. Visual control ratings and height reduction measurements were conducted weekly for four weeks. Dicamba+glufosinate treatments were analyzed to determine if the application resulted in additive, antagonistic, or synergistic control for either weed size or species. Carpetweed was controlled no different between all herbicide treatments and dicamba+glufosinate resulted in additive control. Decreased control between sequential herbicide applications was realized when dicamba was applied first on large crabgrass. Dicamba+glufosinate applied to large crabgrass resulted in additive control. Results of the experiment provide evidence that single and sequential applications of dicamba and glufosinate are efficacious on carpetweed and large crabgrass; however if large crabgrass is a prominent weed in the field being treated, sequential applications were dicamba is applied first may result in reduced control.

IMAGING ANALYSIS TO QUANTIFY LEAF DEFORMATION IN RESPONSE TO SUBLETHAL RATES OF DICAMBA. M Wasacz*, M VanGessel· DJ Mayonado, TE Besancon; Rutgers University, Chatsworth, NJ, University of Delaware, Georgetown, DE, Affiliation Not Specified, Hebron, MD

ABSTRACT

Dicamba is a synthetic auxin herbicide that is prone to off-target movement, including drift and volatilization. Due to the increased acreage of dicamba-tolerant soybean in the mid-Atlantic region to control herbicide-resistant weeds, dicamba drift injury to neighboring vegetable crops is of concern. This method was developed based on a method used by Sassenrath-Cole, et al., 1995 in order to measure leaf deformation caused by dicamba injury. The objective of this method is to quantify crop injury to determine relative sensitivities of several economically important vegetable crops since it is visually difficult to distinguish injury severity between different crop species. This method begins with treating vegetable species of concern with sublethal rates of dicamba to simulate drift conditions. Four weeks after treatment, the first leaf that emerges after treatment, which tends to be foliage with peak injury, is removed. This area of this leaf, called the leaf shadow, is measured in its natural configuration using a leaf area scanner. The same leaf is then flattened, and the area is measured again. The leaf shadow is divided by the flattened leaf area to calculate the Leaf Deformation Index (LDI), which can be used to compare leaf deformation severity within a crop species and across different crop species. Future studies aim to compare several crop species and varieties to establish a registry of vegetable crops and their relative sensitivities that may help refine label recommendations, as well as help growers better design planting strategies around dicamba-treated fields.

EVALUATION OF CHOPPED MISCANTHUS BIOMASS AS A MULCH FOR WEED CONTROL IN CONTAINER GROWN NURSERY CROPS. MH Warren*, JC Neal, CD Harlow; North Carolina State University, Raleigh, NC

ABSTRACT

EVALUATION OF CHOPPED MISCANTHUS BIOMASS AS A MULCH FOR WEED CONTROL IN CONTAINER GROWN NURSERY CROPS. M.H. Warren, J.C. Neal, and C.D. Harlow, North Carolina State University, Raleigh NC ABSTRACT In container production of herbicide-sensitive nursery crops, growers often use mulches, such as parboiled rice hulls, for weed control. However, such mulches are more expensive than preemergence herbicides and a less expensive alternative would be desirable. Chopped miscanthus, (*Miscanthus x giganteus)*, is commercially produced in North Carolina but has not been evaluated for its potential as weed control mulch. The objectives of this study were to compare the efficacy of miscanthus mulch with rice hulls. Specific parameters tested included mulch depth; seed placement; and two cut types of miscanthus. Spotted spurge (*Euphorbia maculata*), was selected as the test species. The experiments were conducted in 2016 at a commercial nursery in Wayne County NC and repeated in 2017 at the NCSU Horticultural Field Laboratory in Raleigh, and in 2020 at a commercial nursery in Johnston County NC. Each year the experimental design was a RCBD with 4 reps and 3 subsamples per plot. Variables compared included 2 mulch types – chopped miscanthus and parboiled rice hulls; 3 mulch depths 0.5, 1.0, or 1.5" (plus no mulch control); and 2 weed seed placements below the mulch or atop the mulch. Percent weed control was visually evaluated, plus weed counts and above-ground fresh weed weights were recorded. Due to differences between the 2016 and 2017 studies, an experiment was conducted in 2020 to compare two cut-types of miscanthus – a more uniform stem-cut (similar to the 2016 product), and a more shredded product (similar to the 2017 product). In each experiment, mulches were less effective when weed seeds were placed beneath the mulch. In 2016, when seeds were placed atop the mulch miscanthus controlled spurge 100% at all mulch depths, not different from rice hulls at ≥ 1 in., but rice hulls controlled only 35% at 0.5" depth. When seeds were placed beneath the mulch, rice hulls controlled spurge better than miscanthus at the 1-inch depth but control was not different at the 0.5 or 1.5 inch depths. Control improved with increasing mulch depth. In contrast, in 2017, 0.5" depth of rice hulls was more effective than 0.5" miscanthus, but there were no differences between mulches at ≥1 in. depth. In the 2020 experiment, chopped miscanthus controlled spurge emerging from below the mulch better than shredded at 1" depth. Both cut types controlled weeds equally well atop at 1 to 1.5" depths. In 2020, 0.5" depth seed placed under the mulches enhanced weed growth, therefore, early hand weeding would be important to stop initial weed development. Based on these experiments, miscanthus mulch was as effective as rice hull mulch for control of spotted spurge when applied at ≥1 in. depth and can be considered as a local, commercially produced economical alternative mulch for weed control in container nurseries saving up to 4 cents per 4-liter pot compared to using rice hull mulch.

EVALUATION OF SUBLETHAL DICAMBA RATES APPLIED TO EARLY AND LATE PLANTED SOYBEAN (GLYCINE MAX). M Fajardo Menjivar*, EA Jones, DJ Contreras, MA Granadino, DE Salazar, W Everman; North Carolina State University, Raleigh, NC

ABSTRACT

Soybean (Glycine max) production locate USA as the leader in terms in production worldwide and the second country with the highest exports of the grain. Animal protein feed and vegetable oil production are the main final processing products of soybeans. Different varieties and maturity groups adapted to the different regions of the country had become in greater yield productions along past years. Dicamba tolerant beans have created the opportunity to farmers to improve weed management and reducing the injury rates of their production fields. Along this the potential of increasing other susceptible crops injury has risen. Defining the different effects of sub lethal Dicamba rates, when exposed at vegetative (V4) or reproductive stages (R2) and with two different planting dates (May or June) in terms of injury and subsequently in yield was the objective of this research. Two different maturity groups (V and VI), each with Dicamba tolerant and non-tolerant soybean varieties were used. The location of the research was the Upper Coastal Plain Research Station at Rocky Mount, NC. The rates of Dicamba to soybeans applied were 0.0, 1.08, 4.35, and 17.40 g ae ha-1. Evaluations were conducted at 14, 21, 28, 35 and 42 days after treatment (DAT). Results provide evidence that there was significant difference of yield between the soybean varieties and rates at a vegetative or reproductive stage.

POSTEMERGENCE HERBICIDES FOR USE IN HEMP GROWN FOR CANNABIDIOL. J Fike, ML Flessner, WC Greene*, KW Bamber, D Reed; Virginia Tech, Blacksburg, VA, Virginia Tech, Blackstone, VA

ABSTRACT

Postemergence herbicides for use in hemp grown for cannabidiolW.C. Greene, M. L. Flessner, K. W. Bamber, J. H. Fike, and T. D. ReedHemp, specifically hemp produced for cannabidiol (CBD), is a new crop to producers in Virginia and little research exists on effective weed control measures, including herbicides. Preliminary research done at Virginia Tech identified several herbicides that may be safe to hemp when applied postemergence. Studies were conducted in Blacksburg and Blackstone, Virginia in order to determine 'BaOx' transplanted hemp tolerance to postemergence herbicides. Herbicide treatments consisted of: 1) fluazifop, (2) clethodim, (3) sethoxydim, (4) bromoxynil, (5) clopyralid, and (6) clethodim + bromoxynil at 10 fl. oz/A, 12 fl. oz/A, 1.5 pt./A, 1 pt./A, 0.33 pt./A, and 16 fl. oz, + 12 fl. oz./A, respectively. Treatments included surfactants according to product labels. A randomized block design with four replications was utilized. Visible injury data as well as at-harvest above-ground biomass was taken. Injury data as well as biomass were pooled across location and subjected to ANOVA and subsequent means separations were performed using Fisher's Protected LSD (a=0.05). Bromoxynil (~4%) resulted in the most visible injury 7 days after treatment (DAT), however by 28 DAT there were no differences in visible injury, which was less than 3% in all treatments. There were no differences in hemp fresh weight at harvest, regardless of herbicide. This research indicates that these herbicides are safe to hemp when applied postemergence, and would be good candidates for receiving a label for use in hemp grown for cannabidiol. Future research will evaluate hemp flowers for herbicide residues.

FLORPYRAUXIFEN-BENZYL APPLIED POST-BLOOM IN APPLES FOR CANADA THISTLE SUPPRESSION. LM Sosnoskie, EC Maloney*; Cornell University, Geneva, NY

ABSTRACT

Orchards in New York are significant contributors to the state's agricultural economy. The New York apple industry is ranked 2nd in the nation, behind Washington, with 55,000 acres in production and providing for more than 10,000 direct jobs. Weeds are a significant problem in orchards and vineyards, especially after transplanting, as well as flowering and early fruit set. Perennial species, including Canada thistle (*Cirsium arvense*, which is a rhizomatous, broadleaf weed in the Asteraceae), are particularly troublesome. In 2020, we conducted a trial at the Cornell research campus in Geneva, NY, to evaluate the effects of florpyrauxifen-benzyl applied alone or in combination with 2,4-D, glufosinate, and glyphosate, post-bloom in apple, on Canada thistle suppression. Herbicides were applied on June 4th, 2020, using a backpack sprayer (2 nozzle boom, Tee Jet 8002VS spaced 19" apart) to plots 24 ft in length and 5 ft in width (three trees per plot). Wind speeds were 2 mph, air temperature was 81 F, and RH was 54%. Canada thistle was 12" to 14" in height and was beginning to bud. Herbicide treatments included: florpyrauxifen-benzyl (1.4 fl oz/A formulated product), glyphosate (48 fl oz/A formulated product), glufosinate (48 fl oz/A formulated product) and 2,4-D (24 oz/A formulated product), applied alone; florpyrauxifen-benzyl (1.4 fl oz/A formulated product) was also applied in combination with the other herbicides). Adjuvants were included as recommended by the labels. Whole plot Canada thistle cover (% of plot area occupied) was assessed weekly and the percent change in cover was determined. Tree injury was not rated because of confounding fire blight damage. At 1 week after treatment, Canada thistle cover was reduced from 0% (glyphosate, alone) to 48% (glufosinate plus florpyrauxifen-benzyl) in the treated plots. Maximum cover reductions, relative to starting cover estimates, were observed at 2 weeks after treatment in the glufosinate (-95%), gluflosinate plus florpyrauxifen-benzyl (-85%), glyphosate (-93%), and glyphosate plus florpyrauxifen-benzyl (-87%) treatments. Florpyrauxifen-benzyl, alone, reduced Canada thistle weed cover by a -45% at 3 weeks after treatment; Canada thistle cover was relatively unchanged in the 2,4-D treatments. At 4 weeks after treatment, Canada thistle cover was reduced by glyphosate -89%, glyphosate plus florpyrauxifen benzyl -84%, glufosinate -54%, and glufosinate plus florpyrauxifen-benzyl -31%. At 4 weeks after treatment, Canada cover increased between +8% and +17% in the 2,4-D, florpyrauxifen-benzyl, and 2,4-D plus florpyrauxifen-benzyl treatments. Ultimately, results show that florpyrauxifen-benzyl, applied alone, was not as effective as glufosinate and glyphosate for short-term Canada thistle suppression. Similarly, florpyrauxifen-benzyl in combination with glufosinate and glyphosate did not improve short-term Canada thistle suppression. Trials will be repeated in 2021 to evaluate rate and timing impacts of florpyrauxifen-benzyl on weed suppression in apples.

SURVEY OF AT-HARVEST WEEDS IN NEW YORK SNAP BEANS. LM Sosnoskie*; Cornell University, Geneva, NY

ABSTRACT

In 2020, we engaged in a multi-year, multi-state project to describe weed escapes at snap bean harvest and describe the environmental and production factors impacting weed control and crop yield throughout the US. The original goal was to survey 30 fields per state at snap bean harvest; a total of 30 quadrats (1 m^2 in size) per field were evaluated where percent (%) weed cover, weed density (numbers per m_2) and snap bean stand establishment (plants per m of row) were rated. Due to COVID restrictions, only 20 fields (in Cattaraugus, Chataqua, Genesee, Livingston, Onondaga, Ontario, Yates Counties) were sampled in 2020. Consequently, 600 total quadrats were surveyed, and 2203 individual weeds identified and counted.A total of 24 species were identified across all fields. Eight species were grasses or grass-like (i.e. yellow nutsedge) and 16 were broadleaves. Four species were perennial and 20 were annuals. Across all quadrats, weed cover ranged from 0% to 95% with an average cover of 11%. Weed density per quadrat ranged from 0 to 59 plants/m^2 with an average density of 4 plants/m^2.The most frequently occurring weed species were:Lambsquarters (*Chenopodium album*), which were found in 28% of quadrats and accounted for 46% (n = 1023) of the weeds that were identified and counted in the study.Crabgrass spp. (*Digitaria spp.*), which were found in 12% of quadrats and accounted for 12% (n = 260) of the weeds that were identified and counted in the study. Pigweed spp. (*Amaranthus spp.*), which were found in 4% of quadrats and accounted for 22% (n = 484) of the weeds that were identified and counted in the study. Other species encountered included: annual bluegrass, barnyardgrass, common chickweed, common ragweed, eastern black nightshade, giant foxtail, green foxtail, hedge bindweed, henbit, horsenettle, oxalis spp. perennial sow thistle, purslane, shepherds purse, velvetleaf, common vetch, volunteer corn and potatoes, wild buckwheat, yellow foxtail, and yellow nutsedge. Hot and dry conditions resulted in poor weed control in certain situations when PRE herbicides were not sufficiently activated and stress reduced POST herbicide efficacy. Grower provided data regarding crop management practices and yields is still being collected. The survey will be repeated in 2021.

SULFENTRAZONE APPLIED PRE-TRANSPLANTING OR POST-DIRECTED FOR WEED CONTROL IN CABBAGE. LM Sosnoskie, EC Maloney*; Cornell University, Geneva, NY

ABSTRACT

NY is the second largest cabbage producing state in the US with more than 23,000 acres. Weeds are a significant problem for growers due to a relatively limited chemical toolbox. New York's cabbage producers have inquired about using sulfentrazone, which is labeled on processing cabbage in other states, including Florida and Michigan, for control of troublesome species, like pigweed. Results from 2019 trials suggested that the active ingredient could be an effective partner to the WSSA 3 herbicides. In 2020, a cabbage herbicide trial was designed to evaluate the efficacay and safety of sulfentrazone (applied pre-transplant (and post-directed). A field site was established at the Cornell research farm in Geneva, New York on May 27th. Sulfentrazone was applied at a rate of 6 oz/A formulated product one day prior to cabbage transplanting (PRE-T) to plots that were 5 feet wide by 25 feet long. A comparison treatment, oxyfluorfen at 8 oz/A formulated product, was also applied to separate plots. Cabbage (two rows/plot on 30" spacing) were transplanted and the plots irrigated to incorporate the herbicides. In a second set of plots, either sulfentrazone or oxyfluorfen were applied at the same rates as a directed spray (DIRECTED) to the base of transplants (June 16th). A three-nozzle boom (Tee Jet 8002 nozzles spaced 19 inches apart) was used to apply the products PRE-T; a one-nozzle boom was used for DIRECTED treatments. Soil and air temperatures ranged from 65 to 80 F, and RH conditions ranged from 47 % to 54%. No weeds were emerged at the time of PRE-T applications; minimal numbers of small weeds (< 2" to 3" in height) were emerged at the time of the directed application. Plant stand (numbers of cabbage/m of row), percent (%) crop injury estimations, percent (%) weed cover estimations, and weed densities (plants/plot) were determined weekly for up to six weeks after transplanting. Cabbage was harvested 75 days after transplanting. Applications of sulfentrazone PRE-T did not reduce mean stand numbers (29 plants per plot) relative to the oxyfluorfen PRE-T treatment (29 plants per plot),nor the untreated check (30 plants per plot) at any time point. Maximum crop injury in response to sulfentrazone and oxyfluorfen, when applied PRE-T, was 8% (for both active ingredients) on June 18th; percent injury ratings fell to less than 1% by July 6th. No appreciable crop injury was observed in response to DIRECTED treatments of either active ingredient. PRE-T applications of sulfentrazone and oxyfluorfen significantly reduced weed cover (which ranged from < 1% (June 5th) to 11% (July 6th)), relative to the untreated check (which ranged from 1% (June 5th) to 91% (July 6th)); there were no statistical differences between the active ingredients with respect to weed control. Broadleaf weed densities (mainly common ragweed) in response to PRE-T sulfentrazone and oxyfluorfen treatments ranged from < 1 plant/plot to 15 plants/plot; broadleaf weed densities in the untreated check ranged from 3 to 41 plants/plot. DIRECTED applications of sulfentrazone and oxyfluorfen reduced late-season weed numbers, relative to the untreated check ,by 62% to 84%. Both active ingredients at both application timings significantly improved mean cabbage weights (2 lbs per head) relative to the untreated check (0.3 lb per head). Results suggest that sulfentrazone applied PRE-T or DIRECTED was safe and effective for use in New York cabbage production.

OPPORTUNITIES FOR USING UAV SPRAYER IN GOLF COURSE NATURAL AREAS. T Reinhardt Piskackova*, A Howell, RG Leon; Czech University of Life Sciences Prague, Prague 6 - Suchdol, Czechia (Czech Republic), North Carolina State University, Raleigh, NC

ABSTRACT

Natural areas of golf courses are an important aspect of golf course ecology that golf patrons value, but can pose a challenge for management. In addition to their terrain often being steeper and harder to access, ground sprayers will often leave track marks in the taller vegetation and cause lodging of native species. UAVs present an opportunity for management in these challenging spaces, but the new technology has not been tested in these areas. Two goals of this research: 1) to evaluate weed control and aesthetics from these two application methods side by side, and 2) to evaluate effectiveness of different carrier volumes. Experiments were carried out at Lonnie Poole Golf Course in Raleigh, NC. Experimental units were 3m by 10m with 3m buffer between. PRE was applied March 17 and POST was applied June 9. Weed cover from the different treatments was evaluated and the time of each application was recorded. A lot of work still needs to be done to optimize drone sprayer systems, but the work seems promising, especially if the low carrier volumes can yield adequate weed control. Testing in short strips can be harder than treating larger areas, so there may be some better experimental design in the future.

A GROWING DEGREE-DAY MODEL TO PREDICT GOOSEGRASS (ELEUSINE INDICA) EMERGENCE IN TURFGRASS. MT Elmore*, M Prorock, DP Tuck; Rutgers University, New Brunswick, NJ, mesur.io, Chapel Hill, NC

ABSTRACT

The objective of this study was to determine if goosegrass (*Eleusine indica*) seedling emergence in cool-season turfgrass is associated with weather variables to develop an emergence model. Goosegrass emergence was assessed weekly in New Jersey, USA at three sites in 2017 and two sites in 2018. Goosegrass seedlings were counted and removed on a weekly basis from April through October within fixed circles. An ensemble modeling approach was used to derive meteorological and agronomic conditions at each site. Growing degree-days were calculated from winter solstice using a 10°C base temperature. Correlation analysis using logistic regression determined week-of-year, growing degree-days, and air temperature were most associated with percent cumulative goosegrass emergence. These variables were subjected to logistic regression using a Gompertz function to describe emergence. Regression analyses that used week-of-year and accumulated growing degree-days described emergence equally well when pooled across locations ($R^2 = 0.74$ and 0.75, respectively). When regression analyses were conducted separately for each location and year, goodness-of-fit values for equations using growing degree-days improved ($R^2 > 0.97$). Emergence patterns were variable among locations and year and this variability was not explained by growing degree-days. This research demonstrates that air temperature-based functions can explain goosegrass emergence. More advanced techniques are being incorporated into an improved model, including a seven-day moving average of air temperature and growing degree-days. This model will be validated across multiple locations in 2021.

EFFICACY OF GRANULAR FLUMIOXAZIN + PRODIAMINE ON WEEDS COMMON TO CONTAINER NURSERIES. JC Neal*, CD Harlow; North Carolina State University, Raleigh, NC

ABSTRACT

Flumioxazin + prodiamine (Fuerte) is labeled for preemergence weed control in container-grown nursery crops. However, at the labeled use-rate of 100 lb/A the dose of flumioxazin applied is 0.125 lb ai/A, about one-third that in the recommended dose of Broadstar 0.25G. It is unclear if the combination of prodiamine @ 0.75 lb ai/A with the lower dose of flumioxazin would provide weed control comparable with industry-standard herbicides. Thus, experiments were conducted to compare Fuerte efficacy with other labeled herbicides, in particular to the recommended dose of flumioxazin applied in Broadstar. Experiments were conducted in 2015, 2018 and 2020 at North Carolina State University. In the 2015 experiment, an experimental formulation of flumioxazin + prodiamine from the Argite Company was evaluated. In 2018 and 2020 the Fuerte formulation from OHP was tested. Treatments included Fuerte (or Argite) at 100 lb/A, Broadstar @ 150 lb/A, Freehand at 150 lb/A, Marengo G at 150 lb/A, Snapshot TG @ 150 lb/A, and Biathlon @ 100 lb/A. Pots were filled with pine bark substrate amended with an industry-standard slow-release fertilizer and hand watered to settle the substrate. Herbicides were applied, then pots were overhead irrigated. Within 24 hours, weeds were surface-seeded. Weed species included spotted spurge, flexuous bittercress, livid amaranth, large crabgrass, rice flatsedge, eclipta, dogfennel, yellow woodsorrel, long-stalked phyllanthus, and chamberbitter. Not all species were included in each year. Woodsorrel, rice flatsedge, dogfennel, chamberbitter, and eclipta were each included in one of three years. When averaged across species, Argite or Fuerte was generally as effective as other labeled herbicides. There was some variability in efficacy among species and between years, with efficacy generally being greater in 2015 and 2020 compared to 2018. The combination of flumioxazin + prodiamine controlled bittercress, livid amaranth, woodsorrel, dogfennnel, and phyllanthus and was not different from Broadstar on these species. Spurge control was more variable. In 2015 the Argite formulation provided better spurge control compared to Broadstar. However, in 2018, Broadstar was better than Fuerte, and there was no difference between Fuerte and Broadstar in 2020. In 2015 Argite controlled crabgrass better than Broadstar, but in 2018 both herbicides were ineffective. In 2020 both herbicides controlled crabgrass $\geq$90%. Both flumioxazin-containing herbicides controlled rice flatsedge 75 to 84% and eclipta 13 to 38%. Overall weed control with Fuerte was comparable to other labeled herbicides, each having some strengths and weaknesses and variable performance between years on some common weed species.

THE EFFECTS OF SOIL MOISTURE CONTENT ON WHEAT (TRITICUM AESTIVUM) AND ITALIAN RYEGRASS (LOLIUM MULTIFLORUM) GROWTH. DE Salazar*, MA Granadino, DJ Contreras, M Fajardo Menjivar, EA Jones, W Everman; North Carolina State University, Raleigh, NC

Abstract

Considered as one of the "big three" cereal crops, wheat (*Triticum aestivum* L.) can yield over 10 ton ha^{-1}, provided sufficient water and nutrients are available, and effective control of pests and pathogens is ensured. Drought stress is one of the major constraints to crop production. There is limited information in wheat and Italian ryegrass response to water deficit in competition. To evaluate competition under moisture stress six water table depth gradient tanks were constructed. A gravel layer at the bottom of the tanks allowed a steady water table beneath the soil which provided a soil moisture gradient. The tanks were split into nine rows, starting at the bottom at approximately 30 cm height from the water table and going up to approximately 155 cm. Decreasing volumetric water content was observed as height from the water table increased. Species composition was made of wheat and Italian ryegrass in a monoculture and in competition. Height reductions due to moisture stress and competition were observed in both species, though greater reductions were observed in Italian ryegrass compared to wheat. Tiller number was reduced in both wheat and Italian ryegrass when grown in competition and when subjected to moisture stress. Tillering was greater in plants grown in a monoculture than when grown in competition, regardless of soil moisture content. Above-ground biomass was greater in Italian ryegrass compared to wheat when not under competition or moisture stress. However, reductions were also greater in Italian ryegrass when in competition with wheat or under moisture stress. Above-ground biomass did not differ between Italian ryegrass grown in competition or in a monoculture when subjected to a water deficit. Wheat above-ground biomass was reduced when in competition with Italian ryegrass and when subjected to a water deficit. Root length increased as moisture content decreased in wheat and Italian ryegrass regardless of being grown in a monoculture or in competition. Above-ground growth was reduced due to competition and moisture stress in both wheat and Italian ryegrass. Though root biomass decreased with decreasing soil moisture content there was an increase in root length in both species regardless of competition indicating a response to moisture stress.

Keywords: Drought, weed interference, moisture gradient, water deficit, moisture stress, competition.

PRELIMINARY SCREENING FOR HERBICIDE TOLERANCE IN TWO NOVEL PERENNIAL GRAIN CROPS. EP Law*, MP Spoth, A DiTommaso; Cornell University, Ithaca, NY, Virginia Tech, Blacksburg, VA, Cornell University, Dryden, NY

ABSTRACT

Perennial grain crops are being developed as alternatives to annual small grains such as wheat due to the potential environmental benefits they can provide. Fewer field operations, particularly tillage, reduces energy and labor inputs, decreases soil disturbance, and protects water quality. Dual-purpose production of both grain and forage from the same crop can also enhance efficiency in the right production system. Over the past four years our work with two perennial grains, ACE-1 Perennial Rye and Kernza Intermediate Wheatgrass, has identified weed management during stand establishment and the post-harvest regrowth period as an important factor in crop persistence and productivity. There are currently no herbicides labeled for grain production for either of these crops, leading us to assess crop tolerance to eight candidate herbicides in a series of three greenhouse trials. All herbicides were applied at label rates for small grains or grass seed production four weeks after seedling emergence. Injury ratings were taken at two, four, and six weeks after application, and shoot height and biomass were measured at six weeks. Injury for each of the herbicides were compared to an untreated control using Dunnett's test. There was minimal visual evidence of herbicide injury during this period, but a few herbicides did cause some stunting. Specifically, Dual II Magnum reduced shoot height of Kernza intermediate wheatgrass by 10.5%, and 2,4-D reduced seedling biomass of ACE-1 perennial rye by 16% and Kernza by 14%. These results demonstrate reasonable crop tolerance to the candidate herbicides, which will allow us to continue with field trials in the near future, but also identified herbicides that might require more specificity in application timing or rate to avoid crop injury.

COMPETITIVE AND ALLELOPATHIC EFFECTS OF TALL FESCUE ON ANNUAL BLUEGRASS.

CG Goncalves*, S Askew; Virginia Tech, Blacksburg, VA

ABSTRACT

Studies have shown that mowing tall fescue at or near four inches will decrease weed populations compared to lower mowing heights. Some researchers have speculated that allelopathy could be involved with competitive displacement of weeds by tall fescue. Greenhouse and growth chamber trials were conducted at Virginia Tech's Glade Road Research Facility in Blacksburg, VA, to 1) examine the impact of three tall fescue mowing heights on annual bluegrass (*Poa annua*) seedling establishment, 2) determine whether soil leachate collected from tall fescue affect germination of annual bluegrass seed, and 3) evaluate the effect of aqueous leaf extracts of tall fescue on annual bluegrass germination and growth. Mowing-height treatments included turf-type tall fescue established from sod maintained at 1.5, 3.0, and 4.5 inches mowing heights, and a soil-only control to determine emergence of annual bluegrass without competition from turfgrass. Soil leachate treatments included 36-inch lysimeters of routinely-fertilized native silt-loam soil containing 'Kentucky 31' tall fescue (*Festuca arundinacea* Schreber), turf-type tall fescue (cv. Falcon), and soil only and a comparison lysimeter that was nonfertilized soil only. The lysimeters were watered daily with 1 quart of solution comprised of collected leachate supplemented with tap water. After 1 month, leachate was used to hydrate seed germination paper along with a distilled water control. Leaf extract treatments included aqueous leaf extracts of 'Kentucky 31' tall fescue, turf-type tall fescue, and distilled water control. Preliminary results suggest that all tall fescue mowing heights result in reducing annual bluegrass germination and growth. Soil leachate did not inhibit germination or seedling growth of annual bluegrass regardless of tall fescue cultivar, soil only, or fertility treatment. Results from leaf extract treatments indicated that different cultivars have different allelopathic effects on annual bluegrass seed germination and seedling growth and that allelopathic activity increases when leaves were obtained from tall fescue plants growing at temperatures below 60 F.

DEVELOPING AN INTEGRATED WEED MANAGEMENT EFFICACY TABLE. ML Flessner*, M VanGessel, KW Bamber, TE Besancon, RS Chandran, T Hines, Q Johnson, D Lingenfelter, CG Rubione, LS Shergill, V Singh, KM Vollmer, JM Wallace; Virginia Tech, Blacksburg, VA, University of Delaware, Georgetown, DE, Rutgers University, Chatsworth, NJ, West Virginia University, Morgantown, WV, Virginia Tech, Painter, VA, Penn State University, University Park, PA, University of Maryland, Queenstown, MD

ABSTRACT

Identifying the best integrated weed management approach can be a daunting task when considering the complexity and multiplicity of weeds, tactics, and integration of these. A poster was developed to simplify integrated weed management in row crops. It summarizes the relative effectiveness of weed management tactics for specific weed types and how best to integrate them. Similar posters or charts are widely available for herbicide selection, but one has not been developed for integrated weed management. The poster contains two tables that rate tactics either before planting (focusing on winter weeds and early emerging weeds) or after planting (focusing on weeds encountered during the crop season). Representative weed species are listed and organized by their characteristics, which include plant type (broadleaf or narrow leaf), lifecycle (annual or perennial), emergence period (early, late, season-long, and similar), growth habit (simple, vining, creeping), and primary reproductive structure (small seed, large seed, tap root, rhizome). Weed control tactics are organized as the crop season progresses and include broad categories of crop rotation, cover crops, tillage and cultivation, planting, fertility, post-harvest, equipment sanitation, and chemical. Within these broad groups, tactics are broken down further. For example, tillage is broken into: no-till, moldboard plow once in four years, annual fall moldboard plow, annual spring moldboard plow, spring disk or chisel plow, and vertical tillage. Relative efficacy of weed control tactics were rated by experts across the mid-Atlantic region on a scale of Detrimental, No effect, Poor, Fair, Good, and Excellent. In some cases, insufficient information is available or tactics are not applicable. The tables allow users to easily and quickly identify and compare efficacy of integrated weed management tactics. The poster is available at https://agweedsci.spes.vt.edu/extension/publications/iwm.html.

EFFECTIVE SOA: AN APP TO FACILITATE SELECTION OF DIVERSE HERBICIDE SITES OF ACTION. ML Flessner*, MV Bagavathiannan, K Pittman, D Hathcoat; Virginia Tech, Blacksburg, VA, Texas A&M University, College Station, TX, Texas A&M AgriLife Research, College Station, TX

ABSTRACT

Applications that include multiple, effective herbicide sites of action (SOA) decrease the risk of herbicide resistance (HR) development and in many cases increases effectiveness. But it can be difficult and time-consuming for farm decision makers to know which herbicides are truly effective on which weeds and how many effective SOA are being applied to each weed. Therefore, we are developing an app to facilitate this process. A herbicide may not be effective due to either tolerance or resistance. To determine HR within the app, a user can input their own HR status, including no known HR, or opt to use the app's internal county-by-county database of HR, which was created from a survey of extension specialists and other experts. Herbicide tolerance is determined in two steps. First, the herbicide product must deliver an effective rate, which is set 75% of the minimum rate of the single active ingredient product. For example, the flumioxazin rate in Envive (flumioxazin + thifensulfuron + chlorimuron) must be at least 75% of the minimum rate for Valor SX (flumioxazin) to be considered effective. If the rate is sufficient, the app consults an internal efficacy table, compiled from various extension guides from across the eastern US. A herbicide is considered effective if its efficacy rating is 80% control or better. App users input information in 5 steps: (1) state and county, (2) crop (corn, soybean, or cotton) with herbicide tolerant trait, (3) weeds (selecting up to 3 of 9 driver weeds in the app), (4) herbicide resistance status (discussed above), and (5) herbicides applied. Within herbicides applied, a user can select up to 5 herbicides within burndown, preemergence, and postemergence. Results are tabulated within application timing and display both the total number of effective SOA and which herbicides are effective for each weed. Users can also link to information on why a herbicide was not effective. The app greatly simplifies identifying effective SOA, which we believe will increase use of best practices to mitigate herbicide resistance.

A23372A - A BROAD-SPECTRUM SOLUTION FOR SUPERIOR WEED MANAGEMENT IN SOYBEAN. C Austin*, B Miller, TH Beckett, P Eure; Syngenta, Greensboro, NC

ABSTRACT

A23372A - A Broad-Spectrum Solution for Superior Weed Management in Soybean.Craig Austin, Brett R. Miller, Tom H. Beckett and Pete Eure, Syngenta Crop Protection. A23372A is a new herbicide being developed by Syngenta Crop Protection for broad-spectrum control of annual grasses and key broadleaf weeds in soybeans. The active ingredients contained in A23372A are S-metolachlor, metribuzin and cloransulam-methyl in a ratio that delivers robust rates of all three herbicides in a convenient mixture. In field testing, A23372A displays excellent crop safety across soil types and environments in all regions of the country. This new herbicide mixture controls annual grasses and most small-seeded broadleaves like waterhemp (Amaranthus rudis) and Palmer amaranth (Amaranthus palmeri) as well as many key larger-seeded weeds including common and giant ragweed (Ambrosia artemisiifolia and trifida), morningglories (Ipomoea) and velvetleaf (Abutilon theophrasti). A23372A is being developed for broad use across all geographies, soil types and tillage systems, and is compatible with common burndown herbicides such as Gramoxone 3.0 SL, glyphosate, 2,4-D and dicamba. A23372A protects soybean yield by providing early season weed management and will provide an excellent preplant or pre-emergence product as the strong residual base for weed management programs regardless of soybean trait platform.

INTEGRATED WATERHEMP MANAGEMENT IN CORN AND SOYBEAN. B Brown*, V Fernandez, M Hunter, J Miller, M Stanyard; New York State Integrated Pest Management Program, Geneva, NY, Bayer, Canandaigua, NY, Cornell Cooperative Extension, Watertown, NY, Cornell Cooperative Extension, Oneida, NY, Cornell Cooperative Extension, Newark, NY

ABSTRACT

Tall waterhemp (*Amaranthus tuberculatus* Moq.) has recently become established in western New York, USA. Initial screenings have suggested herbicide resistance to WSSA groups 2, 5, and 9. To better understand the effective remaining control options, in 2019 and 2020, we implemented replicated small plot trials in corn and soybean fields infested with tall waterhemp. We tested several individual herbicide sites of action, several PRE- or POST-only programs, several two-pass programs, and several programs that integrated herbicides with interseeding and/or row cultivation. Overall, the most effective programs achieved near complete control of the waterhemp through our final sampling in mid-August and were typically two-pass programs that included at least two of the following: herbicides from WSSA groups 4, 14, 15, or 27 or row cultivation. From an integrated weed management perspective, one standout was our corn treatment of a PRE application of a reduced rate of mesotrione (Callisto at 0.4 L/ha) followed by row cultivation and interseeding of annual ryegrass (*Lolium multiflorum* Lam.). Interseeding likely provided little weed suppression in corn but subsequent biomass may reduce fall or spring weed emergence. Separate trials showed that mesotrione was the most compatible residual herbicide with interseeded annual ryegrass.

FRAMING CRITICAL CONTROL WINDOWS FOR WINTER GROWN BRASSICA CARINATA. T Reinhardt Piskackova*, R Tiwari, RG Leon; Czech University of Life Sciences Prague, Prague 6 - Suchdol, Czechia (Czech Republic), University of Florida, Jay, FL, North Carolina State University, Raleigh, NC

ABSTRACT

Brassica carinata L. is a new biofuel crop that is being studied for introduction to cropping rotations in the southeastern United States. Inclusion in existing rotations can offer soil health and agroecological benefits, but farmers will be more likely to adopt it if the weed maintenance is low. In this work we used weed emergence and phenological models in conjunction with the Critical Period of Weed Control (CPWC) to estimate Critical Control Windows (CCW).

ACURON GT LAUNCH - T-MINUS SPRING 2021. LL Smith*, RD Lins, M Kitt; Syngenta, King Ferry, NY, Syngenta Crop Protection, LLC, Rochester, MN, Syngenta, Greensboro, NC

ABSTRACT

Acuron® GT is a new herbicide coming soon from Syngenta for weed control in glyphosate tolerant field corn. Acuron GT will contain S-metolachlor, mesotrione, bicyclopyrone, and glyphosate for postemergence application with knockdown and residual control of grasses and broadleaves. In 2020, field and greenhouse trials were conducted to evaluate Acuron GT for weed control and crop tolerance. Results show that Acuron GT effectively controls many difficult weeds and provides improved residual control and consistency compared to other commercial standards. Acuron GT is not currently registered for use in the US and is not being for sale.

MANAGING HERBICIDE RESISTANT COMMON RAGWEED (AMBROSIA ARTEMISIIFOLIA L.) IN SOYBEAN THROUGH COVER CROPPING AND APPLICATION OF PREEMERGENT HERBICIDE. SM Hirsh*, B Beale, M VanGessel, KM Vollmer; University of Maryland Extension, College Park, MD, University of Maryland, Leonardtown, MD, University of Delaware, Georgetown, DE, University of Maryland, Queenstown, MD

ABSTRACT

Herbicide resistant common ragweed (*Ambrosia artemisiifolia* L.) is prevalent on the Lower Eastern Shore and Southern Maryland. In 2019, common ragweed populations were found to have two or three-way mode-of-action resistance on the Eastern Shore. Early-season management of common ragweed is strongly dependent upon reducing seed germination and controlling ragweed populations prior to soybean planting; therefore this study will evaluate the combination of delaying cover crop termination in order to increase cover crop biomass and competition with weeds, and herbicide control. We performed on-farm trials at three sites with a history of herbicide resistant common ragweed. We investigated ragweed emergence and growth following 1) cover crop termination early-April, early-May, or at soybean planting (mid-May to early-June), with residual herbicide applied either at cover crop termination, at soybean planting, or not at all, and 2) cover crop termination early-May, with no residual herbicide, Command (clomazone), Linex 4L (linuron), Dimetric (metribuzin), Command + Linex, Command + Dimetric, or Linex + Dimetric applied at soybean planting. Experiments were in a randomized complete block design with four replications. Ragweed emerged in early-May in 2019 and mid-May in 2020. When herbicide was applied only at cover crop termination, common ragweed was more prevalent in soybean when cover crops were terminated early-April than when terminated early-May or at soybean planting ($p = 0.006$). There was no decrease in common ragweed prevalence for two herbicide applications (early-April and at soybean planting) compared to one herbicide application at soybean planting. One "at-planting" herbicide application that included residual herbicide provided very good control of common ragweed at all three study sites. Residual herbicides decreased common ragweed prevalence compared to the no-residual herbicide control at two of the three study sites. Soybean yield was not significantly affected by delaying cover crop burndown or using residual herbicides.

SURVEY OF GLUFOSINATE USE IN DIFFERING CROPS AND REGIONS OF NORTH CAROLINA. EA Jones*, CW Cahoon, RG Leon, W Everman; North Carolina State University, Raleigh, NC

ABSTRACT

Surveys were administered at County commodity extension meetings during 2020 to determine how North Carolina farmers are perceiving and using glufosinate for weed control. The survey consisted of ten questions pertaining glufosinate perceptions and use; including what county and commodity was covered at the meeting. The commodities include corn, cotton, and soybean while the counties were pooled by region for a composition of Piedmont, Coastal Plains, Northeast/Blacklands. Out of 343 surveys administered, 296 surveys were completed. Results from the survey provided evidence that approximately half of the surveyed farmers are applying glufosinate for weed control across all regions and commodities. The surveyed farmers applying glufosinate for weed control are doing so for herbicide resistance management. Most glufosinate applications are being applied at the POST (18%) and EPOST+POST (14%) timings. There is a clear indication that North Carolina farmers are aware of glufosinate-resistant weeds evolving as 85% responded as "concerned" of glufosinate resistance as a becoming a problem on their farms. Most farmers have not incurred a control failure with glufosinate (68%), but control failures are more ubiquitous in Piedmont cotton and Coastal Plains soybeans. The weed has that incurred the most control failures when treated with glufosinate is Palmer amaranth (*Amaranthus palmeri*). The responses of the g survey provide evidence that glufosinate perception and use does not differ across commodities or regions of North Carolina. However, control failures with glufosinate applications occur (albeit rare) in certain commodities and regions. North Carolina farmers are aware that glufosinate-resistant weed can evolve and steward this herbicide correctly when the survey was conducted.

INTERSEEDED COVER CROP ESTABLISHMENT AND COMPETITIVE ABILITY IN WIDE ROW CORN GRAIN. TR Mazzone*, JM Wallace; Penn State University, University Park, PA

ABSTRACT

Cover cropping in Pennsylvania grain production systems is limited by the late harvest of grain crops. Producers are interested in the conservation benefits of cover cropping but constrained by narrow cover crop growing season windows when seeding follows grain harvest. Interseeding cover crops is a management practice that enables incorporation of a cover crop into field corn but adoption of this practice has been limited by inconsistent performance. Environmental and management factors that increase competition from the corn crop are likely limiting cover crop establishment and persistence through periods of environmental stress following corn canopy closure. To explore potential limiting factors to interseeded cover crop establishment, a field experiment was initiated in 2020 at RELARC, Pennsylvania Furnace, Centre County, PA and SEAREC, Manheim, Lancaster County, PA. The experiment was a split-split plot RCBD with four replications. Treatments were corn hybrid (determinate variety ZS9598 or a flex-ear hybrid LC0057) and standard versus wide row spacing (76 cm or 152 cm). Plots were split by cover crop interseeding timing (V3 or V6 corn growth stage) and cover crop species (cereal rye (*Secale cereale* L.) sown at 100 kg ha $^{-1}$, annual ryegrass (*Lolium perenne* L. ssp. *Multiflorum* Lam.) at 28 kg ha $^{-1}$, and medium red clover (*Trifolium pratense* L.) at 17 kg ha $^{-1}$).At SEAREC, cover crop biomass increased due to 152 cm corn spacing for all treatments except for cereal rye and medium red clover seeded at V6 in the flex ear corn. Cereal rye seeded at the V6 timing increased in biomass in the 152 cm determinate variety corn. Cover crop timing had no effect on cover crop biomass. Weed biomass was higher in 152 cm corn for all treatments except for cereal rye and medium red clover seeded at V6 in the determinate corn. We observed no differences in weed biomass between V3 and V6 interseeding timing with the exception of the V3 timing resulting in significantly higher weeds in cereal rye plots in 152 cm corn and determinate variety when compared to V6. At SEAREC, cover crop timing had no effect on corn grain yield, but planting corn in wide rows significantly reduced corn yield. The determinate variety yielded higher than the flex ear variety in the 76 cm corn, other treatments did not differ in grain yield. At RELARC, 152 cm corn spacing increased cover crop biomass in annual ryegrass and cereal rye plots and at both cover crop interseeding timing. The exception was cereal rye in the flex ear, v6 timing and annual ryegrass in the determinate variety, v6 timing where no difference in cover crop biomass was observed. No difference was observed in red clover biomass between 152 cm and 76 cm corn. Interseeding earlier at V3 increased both cereal rye and annual ryegrass biomass compared to the later interseeding timing. No differences were observed in weed biomass in annual ryegrass and cereal rye plots except for an increase in weed biomass in 152 cm corn in the V3 timing, determinate variety annual ryegrass. In medium red clover plots, 152 cm corn increased weed biomass at v6 timing in the flex ear variety, and at V3 timing in the determinate variety. At RELARC, 152 cm corn spacing yielded significantly less than standard spacing corn. Cover crop interseeding timing did not impact yields. A significant spacing by corn variety interaction was detected where lower yields were observed in flex-ear treatments at SEAREC in the 76 cm corn but not in the 152 cm corn. Going forward, these corn and cover crop management tactics will be evaluated in the context of alternative residual herbicide programs to refine interseeded cover crop management practices.

MANAGEMENT OF LARGE PALMER AMARANTH PLANTS IN SOYBEAN. M VanGessel*, BA Scott, Q Johnson; University of Delaware, Georgetown, DE

ABSTRACT

Palmer amaranth (*Amaranthus palmeri* S. Watson) has moved into the Northeast region about 10 yrs ago. It is a very competitive plant, capable of cause significant yield losses in summer crops. It has a rapid growth rate with increasing temperatures and thus it is more troublesome in soybeans than corn. Further complicating management is the presence of Groups 2 and 9 herbicide-resistance in the regional populations. As a result, Palmer amaranth is often treated when it is taller than recommended. This study was designed to evaluate various base herbicide programs used in different strategies. Study was a two factor factorial study with base herbicide programs and application strategy as main effects. Base herbicide programs were fomesafen (Reflex or Flexstar), glufosinate, 2,4-D choline, or dicamba. All base programs included glyphosate except the glufosinate programs. Herbicide strategies included a single application, sequential applications, and sequential applications with lactofen. The first application was made when the average Palmer amaranth height was 40 cm and sequential applications were made 7 d later. This study had four site years. Control ratings, 2 wks after application, were less than 83% for all single applications; no treatment differences. Sequential applications improved Palmer amaranth for all base programs. Sequential applications for fomesafen needed to include lactofen for improved control, but none of the other base programs benefited from tankmixtures with lactofen. Season-end biomass for Palmer amaranth was lowest with dicamba and highest for fomesafen (regardless of the formulation). Palmer amaranth seed production was lowest for dicamba and Flexstar formulation of fomesafen. Sequential applications were needed to control large Palmer amaranth plants, but none of the base programs eliminated seed production. The approach for Palmer amaranth with Enlist or Xtend soybeans is similar to conventional, RR, LL soybeans; Enlist and Xtend soybeans do not provide rescue options for Palmer amaranth.

SOYBEAN, PALMER AMARANTH, AND LARGE CRABGRASS GROWTH AND COMPETITION RESPONSE TO MOISTURE STRESS. W Everman*, MA Granadino, DJ Contreras, RG Leon, DE Salazar; North Carolina State University, Raleigh, NC

ABSTRACT

Due to climate change, drought has become an increasing concern to soybean [*Glycine max* (L.) Merr.] farmers. 2012 marked one of the worst droughts since the 1980s causing a plummet in soybean production and increase in soybean prices around the world. Soybean is especially sensitive to drought stress during vegetative stages hastening flowering and physiological maturity, and reducing yield due to a smaller seed size. Drought exposed soybean during flowering stages have been shown to increase pod abortion. Palmer amaranth (*Amaranthus palmeri* S. Watson) and Large crabgrass [*Digitaria sanguinalis* (L.) Scop.] are ranked as two of the most troublesome and common weeds, respectively, in soybean crop production. Palmer amaranth's invasive nature and resistance to certain herbicides have ranked it among the most troublesome weeds in soybean production, reporting losses of up to 78% at a density of only 8 palmer plants m^{-2}. Large crabgrass is a common weed in many cropping systems, though it has been controlled effectively through the use of herbicide, it is capable of reducing soybean yields up to 37% at 16 plants m^{-2}. Limited information on Palmer amaranth and large crabgrass interference in soybean cropping systems under water deficit conditions is available. To evaluate competition under moisture stress 6 water table depth gradient tanks were constructed. A gravel layer at the bottom of the tanks allowed a water table to sit freely and provide a soil moisture gradient. The tanks were split in nine rows starting at the bottom at approximately 30 cm height from the water table and going up to approximately 155 cm. Decreasing volumetric water content was observed as height from the water table increased. Species composition was made up of a monoculture of the three species (Soybean, Palmer amaranth and large crabgrass) and a paired combination of each. Decreasing water content and competition were found to be highly significant factors influencing soybean and large crabgrass growth. Palmer amaranth presented an increased root growth at water deficit, resulting in a high tolerance to moisture stress.

]

2020 SURVEY RESULTS FOR THE MOST COMMON AND TROUBLESOME WEEDS IN GRASS CROPS, PASTURE AND TURF. L Van Wychen*; Weed Science Society of America, Alexandria, VA

ABSTRACT

The 2020 Weed Survey for the U.S. and Canada surveyed the most common and troublesome weeds in the following grass crops: 1) corn (*Zea mays*); 2) rice (*Oryza sativa*); 3) sorghum (*Sorghum bicolor*); 4) turf; 5) pastures, rangeland, or other hay; 6) spring cereal grains; and 7) winter cereal grains. Common weeds refer to the weeds you most frequently see while troublesome weeds are the most difficult to control, but they might not be widespread. There were 317 total survey responses from the U.S. and Canada, of which 90 were from the following Northeastern Weed Science Society (NEWSS) states and provinces: Delaware, Maine, Maryland, Massachusetts, Michigan, New Brunswick, New York, North Carolina, Nova Scotia, Ohio, Ontario, Pennsylvania, Virginia, and Vermont. In the NEWSS region in corn, the top three most common weeds were 1) common lambsquarters (*Chenopodium album*); 2) *Setaria spp.*; and 3) common ragweed (*Ambrosia artemisiifolia*) and the most troublesome weeds were: 1) Palmer amaranth (*Amaranthus palmeri*); 2) common ragweed; and 3) waterhemp (*Amaranthus tuberculatus*). In the NEWSS region in turf, the top three most common weeds were 1) *Digitaria spp.*; 2) dandelion (*Taraxacum officinale*); and 3) white clover (*Trifolium repens*) and the most troublesome weeds were: 1) *Poa spp.*, 2) *Cyperus spp.*; and 3) a tie among three species. In the NEWSS region in pasture, rangeland, and other hay, the top two most common weeds were 1) *Cirsium/Carduus spp.*; and 2) dandelion; and the most troublesome weeds were: 1) *Cirsium/Carduus spp.*; and 2) horsenettle (*Solanum carolinense*). In the NEWSS region in spring cereal grains, the top two most common weeds were 1) common lambsquarters; and 2) common chickweed (*Stellaria media*); and the most troublesome weeds were: 1) *Setaria spp.*; and 2) horseweed (*Erigeron canadensis*). In the NEWSS region in winter cereal grains, the top three most common weeds were 1) common chickweed; 2) *Lamium spp.*; and 3) *Poa spp.*; and the most troublesome weeds were: 1) common chickweed; 2) horseweed; and 3) *Poa spp.*. The top three most common weeds among all grass crops in the NEWSS region were 1) common lambsquarters; 2) *Setaria spp.*; and 3) *Digitaria spp.* The top three most troublesome weeds among all grass crops in the NEWSS region were: 1) *Poa spp.*; 2) horseweed; and 3) *Cirsium/Carduus spp.* The 2020 weed survey data is available at: www.wssa.net/wssa/weed/surveys/.

ACURON XR HERBICIDE - RESIDUAL WEED CONTROL, CROP SAFETY AND YIELD IN CORN. SE Cully, E Hitchner*, TH Beckett, M Kitt; Syngenta Crop Protection, Marion, IL, Syngenta, Elmer, NJ, Syngenta, Greensboro, NC

ABSTRACT

Acuron XR is a new selective herbicide for weed control in field corn, seed corn, popcorn and sweet corn. Acuron XR contains optimized ratios of bicyclopyrone, mesotrione, S-metolachlor and Atrazine that will provide extended residual control of weeds in corn. Field trials were conducted to evaluate Acuron XR for residual weed control compared Acuron and other corn preemergence one pass and two pass products. Results show that Acuron XR will control many difficult weeds in corn and provides consistent, long lasting residual control.

EVALUATION OF PRE/POST HERBICIDES TO MANAGE PALMER AMARANTH IN SOYBEANS.
B Beale*, AW Leslie; University of Maryland, Leonardtown, MD, University of Maryland, College Park, MD

ABSTRACT

Palmer amaranth (*Amaranthus palmeri*) was first confirmed in Southern Maryland crop fields in 2014. Palmer amaranth has spread rapidly across the region posing a significant challenge to crop producers, particularly in soybean systems. Field research conducted in the region during the 2015-2017 seasons evaluated preemergence herbicides and indicated that herbicides with the a.i. flumioxazin or a.i. sulfentrazone provided the most consistent control. In 2018 and 2019, on-farm research trials were conducted coupling the best preemergence treatments with various postemergence treatments containing combinations of dicamba, fomesafen, and/or s-metolachlor applied at an appropriate timing (<10 cm tall plants at 24 DAT) or late timing. (> 10 cm at 42 DAT). A complete block randomized design with four replications was utilized at each location. The treatment area received an early burn down treatment of cover crop in early April using Glyphosate plus 2,4-D. All locations received a burn down application using paraquat prior to planting. All plots including the control received an application of glyphosate at 24 days after planting. Counts of emerged Palmer amaranth plants were taken from the middle 4 rows of each plot, which represented an actual area of 13.9 m^2 (1.524 m by 9.14 m). The number of emerged Palmer amaranth plants was recorded from individual plots starting 10 days after treatment and every 7 days thereafter. Results indicate significantly better control of Palmer amaranth with the use of any pre/post herbicide combination when compared to the non-treated control. In 2018. treatments containing sulfentrazone + metribuzin with no postemergence application resulted in an average of 6.91 plants per m^2 at 52 days after planting. A similar treatment with a late (42 DAT) postemergence application of fomesafen reduced average density to 1.21 plants per m^2 at 52 days after planting. A similar treatment with a timely (24 DAT) postemergence application of either dicamba or fomesafen resulted in complete control with an average density of 0 plants per m^2 at 52 days after planting. In both years, treatments with flumioxazin + pyroxasulfone performed well. In 2018, the residual only treatment of flumioxazin + pyroxasulfone resulted in an average of 0.41 plants per m^2 at 52 days after planting; a residual fb late (42 DAT) postemergence application of fomesafen reduced average density to 0.04 plants per m^2 at 52 days after planting and the treatments with a timely application (24 DAT) postemergence application of either dicamba or fomesafen resulted in complete control with an average density of 0 plants per m^2 at 52 days after planting. Results from trials in 2019 were very similar, although overall density was lower. In both years, all treatments with either flumioxazin or sulfentrazone applied preemergence followed by a timely post at 24 DAT of either dicamba or fomesafen resulted in complete control of Palmer amaranth 8 and 9 weeks after planting.

EVALUATION OF NOVEL HERBICIDE TOLERANT TRAITS IN MD SOYBEAN SYSTEMS. AW Leslie*, B Beale; University of Maryland, College Park, MD, University of Maryland, Leonardtown, MD

ABSTRACT

Herbicide resistant weeds such as Palmer amaranth (*Amaranthus palmeri*) pose significant challenges to Maryland soybean farmers. Previous studies over the past several years have examined various combinations of PRE and POST herbicides using existing herbicide tolerance technology, and have provided farmers guidance in managing this weed species. In recent years, numerous additional herbicide tolerant traits have been introduced, including RR2 Xtend (dicamba, glyphosate); RR2 Xtendflex (dicamba, glyphosate, glufosinate); LibertyLink (glufosinate); LibertyLink GT27 (glufosinate, glyphosate, HPPD); and Enlist E3 (2,4-D, glyphosate, glufosinate). These platforms offer greater flexibility and additional postemergence options that can be incorporated into an integrated weed management program. However, this gives rise to additional questions about the ability to tank mix various contact and systemic products and the effect on weed control and soybean performance. A field experiment was conducted in St. Mary's County, MD in a field infested with glyphosate and ALS resistant Palmer amaranth. A randomized complete block, split-plot design was used to test 15 herbicide treatments and two application timings with four replicates (including a non-treated control). Herbicide treatments included each active ingredient targeting Palmer amaranth (glufosinate, dicamba, 2,4-D) alone, and in combination with sethoxydim and glyphosate to control grass weeds. Treatment protocol included a clean burndown program (paraquat) followed by application of a weak residual (*S*-metolachlor). Soybeans with Enlist and Xtendflex traits were planted, with dicamba and 2,4-D applications separated by an appropriate border. Initial results indicate that all tank-mix combinations tested provided adequate control of emerged Palmer amaranth, with no apparent antagonism, and no effect of application timing. Glufosinate alone provided weaker control of grass weeds than glyphosate or sethoxydim, and there appeared to be an antagonistic interaction between glyphosate and glufosinate in tank-mixes. Future work will repeat the 2020 experiment and include dicamba/glufosinate tank-mixes utilizing newer Xtendflex soybean technology.

DOES CEREAL RYE TOLERATE GRASSY WEED HERBICIDES? D Lingenfelter*, JM Wallace; Penn State University, University Park, PA

ABSTRACT

More farmers are adopting the use of cover crops for various reasons in the Mid-Atlantic region and other cropping systems of the U.S. Cover crops are typically established in the late summer or early fall to provide ground cover during the winter months and for other benefits to the soil and crop management. Cereal rye (*Secale cereale* L.) is a commonly used cover crop species and is one the least expensive seed options. Since cereal rye is a useful cover crop species, some farmers are interested in growing rye to harvest the seed and plant it as a fall cover. This is done to save money on seed costs and out of necessity when there are limited seed supplies. Currently, cereal rye has the fewest number of herbicides registered compared to other small grains; and furthermore, none have activity on grassy weeds. Therefore, field studies were conducted in 2019 and 2020 to determine the safety of various small grain herbicides on cereal rye with an emphasis on those that control weedy grasses. Cereal rye (var. 'Aroostook') was planted each year in late September at 90 lbs/A. Studies were arranged in a randomized complete block design with three replications. Herbicides were applied with a small-plot, CO2-backpack sprayer system that delivered 15 GPA thru TeeJet AIXR110015 nozzles. Treatments were applied to the rye in the fall (2-leaf stage; 2-5 inches tall) and in the spring (9-13 inches tall). Treatments included: flufenacet + metribuzin (0.34 lb ai/A), pendimethalin (0.95 lb), pyroxasulfone (0.106 lb), metribuzin (0.14 lb), mesosulfuron (0.0134 lb), pyroxsulam (0.0163 lb), halauxifen + florasulam (0.0096 lb), thifensulfuron + tribenuron (0.0234 lb), chlorsulfuron + metsulfuron (0.0188 lb, 2020 study only). All herbicides were applied in the fall and spring except for flufenacet + metribuzin, pendimethalin, and pyroxasulfone which were only fall-applied. Appropriate adjuvants were included where necessary. Visual crop injury ratings were collected and anticipated grain yield. Preliminary results for visual crop injury data from the fall application indicate in the 2019 study that all herbicides cause no more than 5% injury, except metribuzin and mesosulfuron (14-56% injury), whereas in the 2020 study, all treatments caused less than 5% injury, except flufenacet + metribuzin, chlorsulfuron + metsulfuron, metribuzin, and mesosulfuron (12-42%). In the 2019 study, spring applications revealed that all treatments caused less than 13% injury. Mesosulfuron and metribuzin caused the most rye injury at both application timings and in both years of the study. However, flufenacet + metribuzin and chlorsulfuron + metsulfuron caused >10% injury with the fall application in 2020. In summary, the preliminary results of these crop safety studies of herbicides currently not labeled for use on cereal rye show there is potential for some to receive registration for this crop, especially those that provide control of weedy grass species such as pyroxsulam. Mesosulfuron showed the highest level of injury (13-42%) when applied either in the fall or spring, however, rye was injured most when applied during the early stages of growth. Metribuzin and flufenacet + metribuzin showed variable injury (2-42%), however more injury was observed from the fall application. Initial data from the 2020 study, reveal that chlorsulfuron + metsulfuron caused 12% injury when applied in the fall, with spring applications still pending. Additional data collection and studies will be necessary before determining which herbicides should be recommended for potential registration in cereal rye.

ASSESSING RESIDUE AND HERBICIDE MANAGEMENT STRATEGIES IN A CEREAL RYE - CORN SYSTEM. TR Mazzone, Z Larson, JM Wallace*; Penn State University, University Park, PA, Penn State, State College, PA

ABSTRACT

The use of cereal rye (*Secale cereale* L.) prior to corn production is increasing in the Northeast due to grower interest in ryelage – corn silage double cropping in dairy-based systems and the soil health benefits of fall sown cover crops in no-till cash grain systems. Development of best management practices (BMPs) for managing cereal rye residue is needed due to potential interactions with corn establishment and growth rates, herbicide fate, weed emergence patterns, and soil water and nutrient cycling that underlies weed – crop competition. A field experiment was initiated in 2019-2020 at Rock Springs, PA to evaluate the effects of alternative cover crop residue and herbicide management tactics on weed population responses and corn yields. The experiment was imposed as a 5 ´ 3 factorial with a split-plot treatment structure. Main plots were cereal rye residue management tactics, including: (1) ryelage, harvested and then terminated with glyphosate 14 days pre-plant (DPP); (2) early kill, terminated 14 DPP; (3) standing, terminated 0 DPP; (4) roll-crimped at planting with integrated ZRX system and double-disk row cleaners; and (5) roll-crimped at planting using a front-mounted I&J style roll-crimper and no row cleaners. Split plots were herbicide management tactics, including: (1) control, glyphosate (1.26 kg ha^{-1}) applied at planting (pre-) and then post-emergence (30 cm corn height); and glyphosate applied pre- and post- with *S*-metolachlor (1.87 kg ha^{-1}) + mesotrione (0.188 kg ha^{-1}) + atrazine (1.82 kg ha^{-1}) applied (2) pre-emergent or (3) post-emergent. Cereal rye was seeded Oct 1 at a rate of 100 kg ha^{-1}. Ryelage was harvest at the late boot stage, which produced 2,680 kg ha^{-1} of dry matter biomass. Delaying termination of cereal rye to 0 DPP resulted in 4,020 kg ha^{-1} of dry matter biomass. Within herbicide control treatments, total weed density did not differ ($p = 0.17$) among residue management treatments. However, in-row weed density was significantly higher in roll-crimped (ZRX) treatments that utilized double disk row cleaners compared to alternative late terminated (0 DPP) treatments. Between-row weed density was significantly lower in rolled treatments compared to ryelage and early-kill treatments. Horseweed density (*Erigeron canadensis*, L.) was significantly lower in rolled treatments compared to early kill treatments. A significant residue ´herbicide interaction was observed in analysis of late season weed biomass. Within ryelage and early-kill treatments, use of a herbicide mixture at the post-emergence timing decreased total weed biomass compared to use at the pre-emergent timing, with both timings resulting in lower weed biomass than the control. In late-terminated (0 DPP) treatments, use of the herbicide mixture at the post-emergence timing resulted in lower weed biomass than other herbicide treatments, which did not differ. Finally, corn yield was significantly lower in the ryelage treatment compared to other residue management tactics, which did not differ. This field experiment will be replicated in 2021-2022 with complementary on-farm strip trials to inform the development of cereal rye residue and herbicide management BMPs.

EFFICACY OF DRONES FOR IN-SEASON FOLIAR APPLICATION OF PESTICIDES IN CORN AND SOYBEAN.
AA Kness*, EN Crowl; University of Maryland Extension, Street, MD, University of Maryland Extension, Cockeysville, MD

ABSTRACT

Drones are becoming increasingly popular in agriculture for not only imagery, but product application. As a result, startup companies offering aerial pesticide application via drone are emerging. Farmers have taken interest in the technology and service for several reasons, but the main benefit to using a drone to apply crop production products in soybean and corn is that it offers a feasible method for in-season foliar product application to fields that are smaller, fragmented, or irregularly shaped, without the potential for damaging the crop with a ground spray rig. Additionally, drones may have an advantage over helicopters or fixed-winged aircraft in small fields because they are more nimble and have the potential to achieve application to field edges that would be missed by aircraft. Finally, drones are much less intrusive to curious neighbors who often raise concerns when they see an aircraft applying products to fields. Although drones offer a lot of potential, there is very little published data on their efficacy to apply products, which is cause for question and concern as to if drones are a viable and worthwhile means of applying products, such as pesticides, to corn and soybeans. Additionally, drones tend to lack spray tank capacity, so spray volumes with drone applications are low (1-2 gallons per acre). These low-volumes may pose challenges in achieving adequate coverage; however, products are applied at a greater concentration. This research project, funded by the Maryland Soybean Board, aims to collect data regarding drone spray efficacy in corn and soybean.We placed water-activated spray cards at three heights in the plant canopy in R6 corn and R4 soybean. The plots were then sprayed with two different drones (DJI and HSE-TTA). Cards were retrieved from the plots and scanned into a computer. The images were analyzed for droplet coverage, droplet density, and droplet size using the program DepositScan.Averaged across both drones and both crops, the drones achieved an average spray coverage percentage of 2%, droplet density of 20.5 droplets/cm^2, with average droplet size of 150 μm in the top canopy. These results indicate that drones have the potential to apply pesticides and other crop protection products to corn and soybeans, but more work should be done to identify methods to achieve greater spray coverage.

CRITICAL PERIOD FOR ITALIAN RYEGRASS (LOLIUM MULTIFLORUM) CONTROL IN WHEAT (TRITICUM AESTIVUM) IN NORTH CAROLINA. DJ Contreras*, W Everman, EA Jones, M Fajardo Menjivar, DE Salazar; North Carolina State University, Raleigh, NC

ABSTRACT

Field studies were conducted in North Carolina to determine the critical period of weed control (CPWC) for *Lolium multiflorum* L. (Italian ryegrass) in *Triticum aestivum* L. (wheat) with an acceptable yield loss of 5%. Wheat was planted in late fall in 2017 and 2018 in no-till fields. Treatments consisted of allowing weeds to grow from crop emergence for different intervals until removal ("weedy"), and to maintain "weed-free" conditions from crop emergence for the same intervals, and then let the weeds emerge and compete with the crop. In 2017, weed control or removal as done in two-week intervals up to 18 weeks after crop emergence (WAE) and three-week intervals up to 18 WAE in 2018. Treatments were compared through a regression analysis where timing of weed removal was related to yield loss. Additional biological measurements including Italian ryegrass density, and height were collected at 6, 12, and 18 WAE to understand the effect of crop-weed interactions on the CPWC and weed populations. The CPWC for Italian ryegrass in no-till planted wheat was estimated from 1200 – 1860 GDD.

RECENT EXTREME WEATHER EVENTS: ANIMAL DEATHS LINKED TO TOXIC PLANT AND PEST OUTBREAKS. MA Bravo*; Rutgers University, Dept. of Agriculture and Natural Resources, New Brunswick, NJ

ABSTRACT

Centuries of research and post-mortem findings document poisonous plants, plant-pest, plant-bacterial, plant-pesticides and accumulating toxic levels of soil elements in forage and grain in all stages of harvest, either by grazing, or through consumption of processed feed, or act of accessing water, as the cause of sudden death or chronic sickness in livestock. Domesticated bees (*Apis melifera* L.) are livestock, and like other domesticated herbivores, dependent on both cultivated and naturally occurring plants for food sources. In the last decade, droughts, wildfires, hurricanes, floods and exceptionally long-lived derechos and tornadoes resulted in thousands of livestock and forage indemnifications due to poisonous plant toxins. The alarming number of losses in the nation's bee colonies is an example of this trend. This year marks the sixth that ten or more consecutive extreme weather events occurred in the United States. Duration and frequency of multiple consecutive billion-dollar extreme weather events in the same and consecutive growing season raise the following questions: 1. Are we sufficiently prepared and trained to recognize impending plant toxicity scenarios? 2. Are we prepared to act quickly to mitigate forage and feed poisonings before losses occur? 3. Can we deploy rapidly to relocate affected herds to mitigate further harm? 4. Do we have a mechanism to nationally communicate frequency, severity, duration and overlap of multiple extreme weather related plant poisonings in grazed forges, harvested forages and processed feed? 5. Are we appropriately accessing the livestock groups at risk? 6. Are we documenting the changing levels of these toxins in forages and feed in real time? Applied herbicides, insecticides, fertilizers, and manure can raise, lower, accumulate, or not affect poisonous concentrations in plants. Naturally occurring soil elements are also culpable in extreme weather-related livestock mortalities. Outbreaks of mycotoxin, bacterial, viral and pest-contaminated forages and feed kill thousands of livestock worldwide. Toxic plants include non-vascular bio accumulating macrophytes ingested when consuming shrinking water supplies and four dozen other families with toxic members found in lower vascular, gymnosperm, flowering, and monocot divisions. The 2018 Farm Bill language now includes vector and vector-transmitted losses as eligible livestock, forage, honeybee, fish and forestry losses due to extreme weather events. Key plants play a critical role in the life cycle of livestock-killing flies, mosquitos, ticks and other plant-dependent deleterious pests swarming during and after extreme weather events. Blister beetle (*Epicauta* L.) infested hay that killed horses in Wisconsin this year is one example. Historically, botanists, extension professionals, and plant, pest, and soil specialists assist with investigations into livestock deaths, but only veterinarians can certify cause of death during and after extreme weather events for federal loss programs. There is currently a shortage of livestock veterinarians and need for a nationally coordinated poisonous plant toxin and toxin in feed (PPTTF) alert to deploy PPS specialists to assist with emerging toxic plant-pest events and need for daily broadcasted plant-pest weather alerts until the levels of toxin in the forage, feed or water is safe for livestock consumption.

HERBICIDE EFFICACY FOR TERMINATION OF BRASSICA SPP. AND CONTROL OF ITS VOLUNTEERS IN CORN. V Singh*, T Hines, ML Flessner, KW Bamber, M Reiter; Virginia Tech, Painter, VA, Virginia Tech, Blacksburg, VA

ABSTRACT

Herbicides have been the most preferred tools to terminate cover crops before they begin reproductive growth. Despite the benefit of greater biomass at late termination, it may result in seed production which ultimately leads to greater infestations of volunteer cover crop plants in cash crops during summer and after crop harvest in fall. These volunteer plants act as weeds and lead to mixed cover crop stand. Studies were conducted at Painter and Blacksburg, VA, with three major objectives to address the issue of termination timing of cover crops, limited herbicide options and volunteerism. Four cover crops (wheat, cereal rye, mustard, and hairy vetch) were planted in first week of November 2019 and were terminated at different timings (first week of April, Mid-April, and first week of May 2020, and no-termination). Mustard and hairy vetch were terminated with glufosinate+2,4-D, and wheat and cereal rye were terminated with glyphosate. These cover crops were followed by corn (planted first week of May 2020). Results indicated that efficacy of glufosinate was little lower when mustard was terminated during first week of April compared to late termination during first week of May 2020. Late termination and no-termination treatments resulted in similar biomass production, however, early terminated cover crops produced lower biomass compared to other termination treatments. Hairy vetch produced lowest biomass compared to other cover crops at early termination. Non-termination of cover crops (only rolled) led to lower corn yield. Non-terminated (rolled) mustard reduced corn yield by 86%. In terms of weed control, wheat and cereal rye performed better compared with mustard and hairy vetch. Studies will be repeated in 2021 with additional objectives and treatments where 20 preemergence and postemergence herbicides will be evaluated for the control of volunteer brassica spp.

WHAT'S NEW IN INDUSTRY. SA Mathew*, P Eure, DL Bowers, M Kitt; Syngenta, Germantown, MD, Syngenta, Greensboro, NC

ABSTRACT

[no abstract entered]

NYS AGRICULTURAL WEED ID NETWORK: CONNECTIONS AND RESOURCES. A DiTommaso, C Marschner*, S Morris; Cornell University, Dryden, NY, Cornell University, Ithaca, NY

ABSTRACT

The NYS Agricultural Weed Identification Network was founded in 2018. We have built a network of extension collaborators from around the state, established lists of problem and common weeds for the major categories of commodities grown in New York, and created a website of resources for growers in New York and beyond. We have new tools for spring rosette-forming weed identification and weedy mustards, and have delved into poisonous plants for livestock and building a catalog of species profiles for common New York weeds. In the remaining year of the program, we hope to incorporate new tools for grass ID and other knotty identification problems, create an online training video for weed seedling ID, and continue to supply ID services for growers and extensionists.

RENOVATING TURFGRASS WITH ALTERNATIVES TO GLYPHOSATE AND EFFECTS ON SUBSEQUENT WHITE CLOVER AND TALL FESCUE ESTABLISHMENT. D Koo*, JM Peppers, S Askew; Virginia Tech, Blacksburg, VA

ABSTRACT

Turfgrass renovation is a method to reestablish the turfgrass after removing existing deteriorated turfgrass surface. Glyphosate (Roundup) is the most widely used herbicide to burndown the existing turfgrass. Since the recent controversies on the glyphosate, there is an increasing chance of losing glyphosate as viable options for turfgrass renovation. Therefore, various glyphosate alternatives were tested in the respect of control of a mixed vegetative sward, desirable vegetation reestablishment, and cost effectiveness. Field trials were conducted from September to December 2020 at Virginia Tech's Glade Road Research Facility in Blacksburg, VA. 6”x 6” plots were prepared. Glyphosate was applied to 2” x 6” on the center of the plots and glyphosate alternatives were applied to the whole plots at the following week. Tall fescue and white clover were planted 1, 2, and 4 weeks after herbicide application on the 2” x 6” dimension previously burn-downed with glyphosate. All herbicide treatments referred to the maximum rate on the specimen label. All treatments were replicated four times and herbicides were applied with a CO_2-powered boom sprayer calibrated to deliver 40 gal/ac. (374 L/ha) Visual assessment of green cover (%) of tall fescue and white clover were conducted and green cover (%) of targeted weedy vegetation after herbicide application were assessed. The result indicates that dazomet (Basamid), glufosinate with fluazifop-P-butyl (Finale+Ornamec), glufosinate with pinoxaden (Finale+Axial XL), and glufosinate with diquat dibromide (Finale+Reward) shows similar green cover of targeted weedy vegetation with glyphosate. Among these treatments, there are no significant differences on the green cover (%) of tall fescue and white clover seeded in 1, 2, and 4 weeks after treatment. Economic analysis was conducted based on the application rate. Glyphosate ($31.6/A) is about 4 times cost efficient than glufosinate with pinoxaden ($124.31/A) which was the lowest among the alternatives with identical efficacy with glyphosate.

MULTIYEAR PROGRAMS FOR BERMUDAGRASS CONTROL IN CREEPING BENTGRASS FAIRWAYS. JM Peppers*, S Askew, J Brewer; Virginia Tech, Blacksburg, VA

ABSTRACT

Bermudagrass (*Cynodon dactylon* L.) is a persistent and troublesome weed in creeping bentgrass (*Agrostis stolonifera* L.) fairways throughout the transition zone. The persistence of bermudagrass in creeping bentgrass fairways reduces turf quality due to differences in growth habit, leaf texture, and color. Selectively controlling bermudagrass in creeping bentgrass is challenging due to the majority of herbicides that control bermudagrass are often also injurious to creeping bentgrass. Previous research suggests that proper timing can maximize bermudagrass control in the fall when creeping bentgrass is actively growing and bermudagrass is going dormant. Preliminary research has indicated that mechanical slicing of bermudagrass in combination with herbicide applications in the fall can increase bermudagrass control in creeping bentgrass. The objective of this research was to compare a topramezone-based herbicide program alone to a topramezone-based herbicide program plus mechanical slicing for bermudagrass control in a creeping bentgrass fairway. We hypothesize that slicing plus herbicide applications will reduce bermudagrass coverage more than herbicide applications alone. A multiyear study was conducted from September 2018 to September 2020 to determine if the addition of slicing to a topramezone-based herbicide program would increase bermudagrass control compared to the herbicide program alone. The study was arranged as a factorial randomized complete block design with four treatments. Treatments included were mechanical slicing alone, topramezone plus triclopyr applied at 6.4 and 26 g ai ha^{-1}, respectively, and a combination of slicing and topramezone plus triclopyr. Six applications per year of herbicides and slicing were made with three occurring in the fall and three occurring in the spring. Bermudagrass and creeping bentgrass coverage were rated visually over the course of the study. Bermudagrass and creeping bentgrass coverage were converted to area under the progress curve per day (AUPC) to account for repeated measures and to statistically compare the treatments in this study. Treatments that included herbicide applications had less AUPC for bermudagrass coverage than the nontreated and slicing only. There were no significant differences in bermudagrass coverage between the nontreated and slicing only treatments. Although there were numerical differences that suggested that the slicing plus herbicide treatments reduced bermudagrass coverage greater than herbicide treatments alone, no statistical differences were observed in this study. From this study we can conclude that slicing alone is insufficient to reduce bermudagrass coverage in a creeping bentgrass fairway. Topramezone plus triclopyr programs significantly increase creeping bentgrass and decrease bermudagrass coverage and the addition of slicing does not decrease bermudagrass coverage compared to herbicide alone.

HERBICIDE APPLICATION TIMING AFFECTS DEER-TONGUE GRASS (DICHANTHELIUM CLANDESTINUM) CONTROL IN NO-MOW FINE FESCUE. KH Diehl*, MT Elmore, DP Tuck; Rutgers University, New Brunswick, NJ

ABSTRACT

Weed control for no-mow, or "naturalized areas" on golf courses can be difficult and often requires extensive manual labor. The objective of this research was to evaluate the efficacy of fluazifop and glyphosate at various application timings for deer-tongue grass control in naturalized areas. Research was conducted in 2020 from April to October at the Mendham Golf and Tennis Club (Mendham, NJ). Treatments were arranged in a two-by-five factorial, with fluazifop (280 g ha-1; with NIS 0.25% v/v) and glyphosate (560 g ha-1) applied singly at five different application timings. Treatments were replicated five times and arranged in a RCBD with 2.0 by 3.0 m plots. Herbicides were applied using a CO_2-powered sprayer and four-nozzle boom equipped with 1103VS nozzles (TeeJet AIXR) and with a carrier volume of 410 L ha^{-1}. Application timings were selected using a combination of growing degree-days (GDD; base 10° C), cooling degree-days (CDD; base 20°C), and weed developmental stages visually evaluated on site. Each herbicide was applied singly at 75 and 175 GDD (28 April and 26 May 2020), during spring flowering (June 18), in mid-July (July 22), and at 25 CDD (September 22). Plots were mowed once in October, two weeks after the final treatments were applied. To determine percent deer-tongue grass control, deer-tongue grass injury was visually assessed on a 0 (no control) to 100 (complete necrosis) scale every two to three weeks from May through October. Deertongue cover was also visually evaluated from 0 (no cover) to 100 (complete canopy cover) in each plot. All data were subject to ANOVA as a factorial using the GLIMMIX procedure (P=0.05) in SAS (v. 9.4). The effect of herbicide treatment on deer-tongue grass control was significant on each rating date from 3 to 5 WAT and application timing was significant from 9 to 15 WAT, and at 21 WAT. A herbicide-by-application timing interaction for deer-tongue grass control was detected from 14 to 21 WAT and trends were similar on each date. Glyphosate applied at 175 GDD and spring flowering provided greater control (89 and 97%) than glyphosate applied at 75 GDD and 25 CDD (36 and 62%) and all fluazifop treatments (11 to 59%) on 22 October 2020. Glyphosate applied at 25 CDD (22 September) and glyphosate and fluazifop applied mid-July provided 59 to 79% deer-tongue control by the final rating in October. Both herbicides were least effective when applied at 75 GDD (26 April), with glyphosate and fluazifop providing 36 and 11% control, respectively at the final rating in October.

EFFECTS OF PHOSPHORUS LEVELS ON ANNUAL BLUEGRASS ENCROACHMENT IN CREEPING BENTGRASS DURING ESTABLISHMENT. BC McNally*, MT Elmore, AR Kowalewski, AB Cain, ET Braithwaite, JA Murphy, SL Murphy; Rutgers University, New Brunswick, NJ, Oregon State University, Corvallis, OR

ABSTRACT

Annual bluegrass (ABG; *Poa annua* L.) is a problematic winter annual in turfgrass, particularly during seeded establishment. Phosphorus influences annual bluegrass as growth, establishment, and infestation is favored in high P soils. The objective of this research was to evaluate the effect of phosphorus application rate on sward composition when annual bluegrass and creeping bentgrass (CBG; *Agrostis stolonifera* L.) were sown in competition. Research was conducted at the New Jersey Agricultural Experiment Station in New Brunswick, NJ in a glasshouse. Treatments consisted of five phosphorus rates (0, 12.5, 25, 50, 100 kg P ha^{-1}), which were slightly adjusted in run 2 (0, 6.25, 12.5, 25, 50 kg P ha^{-1}). The growing medium was a silica sand and peat moss mix (90:10 v/v) with a P concentration of 3 ppm. Annual bluegrass and creeping bentgrass were sown in 25 cm diameter pots at a seeding rate of 55 and 165 seeds pot^{-1}, respectively. Pots were arranged in a randomized complete block design with four replications. Line-intersect grid counts (49 intersects, 15mm increments) were collected monthly to determine coverage of annual bluegrass, creeping bentgrass, and bare soil in each pot. Percent green cover was assessed monthly using digital image analysis on photos taken from a lightbox. Visual assessments of turfgrass quality (TQ) were conducted at 8 and 11 weeks after seeding (WAS) using a 1 to 9 scale where 1 equaled low-quality turf, 9 equaled high-quality turf, and 6 was considered the minimum acceptable level of turf quality. Aboveground dry shoot biomass was measured at the end of the experiment. Phosphorus influenced annual bluegrass, creeping bentgrass, and bare soil coverage at 8 and 11 WAS. Percent bare soil determined by grid counts was strongly correlated with percent green cover as determined by digital image analysis at 8 and 11 WAS ($r = -0.87$ and -0.98, respectively). Percent bare soil was greater for treatments that received 0 kg P ha^{-1} (>70%) than treatments that received phosphorus (<10%). Differences in annual bluegrass and creeping bentgrass cover were only detected 11 and 8 WAS, in runs 1 and 2, respectively among non-zero P rates. In run 1, 100 kg P ha^{-1} resulted in less annual bluegrass (50%) than 25 kg P ha^{-1} (59%) and more creeping bentgrass (49 and 40%, respectively). In run 2, P applied at = 12.5 kg ha^{-1} resulted in more annual bluegrass (37 to 44%) than 6.25 kg P ha^{-1} (30%). Turf quality was poor in 0 kg P ha^{-1} treatments (<2.9), but all treatments that received P had similar turf quality (6.9 to 7.2) at 8 WAS in run 1. Only the 0 kg P ha^{-1} resulted in less shoot biomass (0.17 g) and there were no differences among treatments that received P (6.59 to 7.84 g) in run 1. Annual bluegrass and creeping bentgrass cover was not substantially affected when P rates as low as 6.25 kg P ha^{-1} were compared to higher P rates (50 kg ha^{-1}) when the two species were seeded together. It is unclear whether there is a soil P concentration that results in a competitive advantage for annual bluegrass during seeded establishment.

A META ANALYSIS OF NORTH CAROLINA WEATHER CONDITIONS AND THEIR IMPACT ON PESTICIDE APPLICATIONS. M Fajardo Menjivar*, EA Jones, DJ Contreras, MA Granadino, DE Salazar, CW Cahoon, W Everman; North Carolina State University, Raleigh, NC

ABSTRACT

In 1962 with the publication of the book “Silent Spring”, written by Rachel Carson, the detrimental effects pesticides may bring to the environment were publicly highlighted. Herbicides use has seen an increase over the past 50 years because of different technologies that have helped increase yield production. Pesticide drift has been an issue and studies have been conducted to investigate the effect of off-target depositions. With the objective of analyzing if weather conditions based on difference of temperature between 2 m and 9 m above ground level and wind speed and gusts are suitable for spraying pesticides and how that changes along the different months of a growing season this research was conducted. Probability of Detection (POD) and Bias were calculated as they help to perform weather analytics to determine conditions. And with them calculated it allowed to interpret how temperature inversions can be detected in the different months of a growing season.

ORGANIC NO-TILL SOYBEAN SEEDING RATE AND NITROGEN FERTILIZATION EFFECTS ON WEED SUPPRESSION. UD Menalled*, MR Ryan, SJ Pethybridge; Cornell University, Ithaca, NY, Affiliation Not Specified, Ithaca, NY

ABSTRACT

Organic farmers rely on soil tillage and cultivation for weed management, which is labor and fuel intensive and can degrade soil health. No-till planting into rolled-crimped cover crops is an alternative practice that can promote soil health and reduce labor and fuel inputs compared with tillage-based farming. Little information is available to help farmers manage organic no-till planted crops. Weed suppression and crop yield were assessed across soybean (*Glycine max* (L.) Merr.) planting rates and soil nitrogen states. Soybeans were no-till planted into rolled-crimped cereal rye (*Secale cereale* L.) at 0, 185,000, 371,000, 556,000, and 741,000 seeds ha^{-1}. To understand the effects of nitrogen on soybean growth and weed suppression, three rates (0, 63, or 125 kg ha^{-1}) of sodium nitrate (15-0-2) were applied across each soybean planting rate main plot. Weed biomass was lowest at the high soybean planting rates, and it increased with nitrogen. Soybean yield increased asymptotically with crop density and was not affected by soil nitrogen.

CEREAL RYE SEEDING RATE EFFECTS ON WEED SUPPRESSION IN PLANTING GREEN SOYBEANS. TS Ficks*, JM Wallace, D Lingenfelter; Penn State University, University Park, PA

ABSTRACT

The delay of cover crop termination by using planting green practices has increased in grower adoption. However, cereal rye (*Secale cereale* L.) seeding rates vary considerably when planting green practices are used. Better understanding of the relationship between cereal rye seeding rates commonly used by growers and winter and summer annual weed suppression while using planting green practices is needed to inform development of best management practices. We conducted a field experiment using a RCBD with 4 replicates to evaluate weed responses to four cereal rye seeding rate treatments: 0, 51, 101, and 135 kg ha^{-1}. The study was replicated in 6 site-years across the 2018-2019 and 2019-2020 growing season and four locations: Rock Springs, PA, Landisville, PA, Georgetown, DE, and Blacksburg, VA. Dicamba-tolerant soybeans (*Glycine max*) were roll-planted using a no-till planter equipped with ZRX integrated rollers. Glyphosate and dicamba were applied for cover crop termination 1 day after planting and for post-emergent weed control at the V4 soybean growth stage. Rye and winter annual weed biomass and horseweed (*Erigeron canadensis*) density and heights were collected approximately 1 week prior to soybean planting. Horseweed and summer annual weed densities and heights were collected at the post-emergent herbicide application timing. Peak season weed biomass was collected in mid-August and soybean yield was taken in October or November. Linear regression and generalized linear mixed-effect models were used in our statistical analysis. Separation of means were conducted at $p<0.05$ with a Tukey's HSD. Only PA site years ($n = 3$) are reported in these results. When the control treatment (0 kg ha^{-1}) was excluded, the linear relationship between cereal rye seeding rate (51 – 135 kg ha^{-1}) and rye biomass, as well as each weed response variable, was not significant. In analysis of cereal rye seeding rates relative to the untreated control, we found that each seeding rate significantly reduced winter annual weed biomass, horseweed density, and summer annual weed density when compared to control treatment. Peak season weed biomass and soybean yield did not differ across seeding rate treatments. Our PA results suggest that increasing cereal rye seeding rates does not increase the magnitude of the weed suppression benefit but has potential to reduce input costs associated with weed management. Data from the Delaware and Virginia site years will be analyzed to further understand these effects across a range of soil-environment conditions.

STRATEGIES FOR USING VARIABLE PLANTING FOR INCREASED WEED SUPPRESSION. SA Ramsey*, DL Jordan, W Everman, A Locke, RG Leon; North Carolina State University, Raleigh, NC, USDA, Raleigh, NC

ABSTRACT

There is a need for cultural and mechanical practices in row crop agriculture to control herbicide-resistant (HR) weeds. An important limitation for developing effective integrated weed management, especially with HR populations, is the lack of strategies to prevent or at least reduce the production of new seed by weeds surviving herbicide application or cultivation. New precision planters with GPS control, currently being acquired by growers, enables the use of variable planting for site-specific weed suppression to limit reproduction of weeds escaping conventional control without increasing costs. Increasing seeding rates could increase competitiveness of the crop with weeds but also decrease yield. Conversely, decreasing seeding rates could increase or maintain yield in weed free areas. The objective of this study was to identify high density planting arrangements in cotton that would favor weed suppression and protect crop yield under weedy conditions and low density arrangements that increase relative crop yield under weed free conditions. Through the use of the solar corridor concept, we created four different planting arrangements in order to maximize light interception to lower layers within the crop canopy and increase crop yield. The results show that low density plantings were able to increase or maintain yield and high density plantings did not decrease yield. The results of the present study will allow the development of a bioeconomic model to optimize high and low planting density allocations based on weed mapping.

SOYBEAN PERFORMANCE AFTER CHAFF LINING AT WHEAT HARVEST. MP Spoth*, ML Flessner, V Singh; Virginia Tech, Blacksburg, VA, Virginia Tech, Painter, VA

ABSTRACT

Chaff lining is a form of harvest weed seed control (HWSC) in which the chaff fraction of crop residues and weed seeds harvested with the crop are concentrated into a narrow line (35 – 40 cm wide) behind the combine. Chaff lining has proven effective and economical in Australian wheat production but has not been evaluated in U.S. wheat/soybean double crop systems. As crop yields increase, so does the amount of chaff in the chaff line, which may impact subsequent crop establishment and weed control. The objectives of this study were to determine how varying wheat yields will affect 1) chaff lining control of two problematic weed species: Italian ryegrass (Lolium perenne L. ssp multiflorum (Lam.) Husnot) and wild mustard (Sinapis arvensis L.), and 2) examine the effect of wheat chaff lining on subsequent soybean crop performance. Separate experiments were conducted for each weed. Each experiment was conducted across three locations in Virginia with four replications per treatment in a randomized complete block design. Chaff lines were created to mimic low=1680 kg/ha (25 bu/a), medium low=3360 kg/ha (50 bu/a), medium high=5045 kg/ha (75 bu/a), and high 6725 kg/ha (100 bu/a) wheat yield treatments. In these treatments, weed seed amounting to 80% of total fecundity, representing weed seeds retained at harvest and harvested with the crop were mixed into the chaff line. Two control treatments were also included: a conventional harvest where 100% of total weed fecundity were evenly broadcasted and an outside a chaff line treatment where 20% of weed seeds were evenly broadcasted, representing weed seed rain prior to harvest. Data were pooled across location and subjected to ANOVA followed by means separation with Fisher's protected LSD(0.05) or linear regression as appropriate. Our preliminary results from one year of data indicate there was a 87.7% reduction in soybean emergence through chaff lines when compared to conventional harvest. No differences were observed in soybean yield across treatments on a field scale, which accounts for both chaff lined and outside the chaff line areas, compared to conventional harvest. Total weed emergence over the growing season was reduced by 53.5% for Italian ryegrass and 76.8% for wild mustard in chaff lined plots compared to the conventional harvest treatment, indicating chaff lines may alter the environment to be unfavorable for weed seed survival or emergence. This reduction in weed density in chaff lined versus conventional harvest could explain the observed soybean yield response. Weed survival was marginally explained by varying wheat yield/chaff amount in wild mustard (R2 = 0.14, Y = 963 - 5.6*Yield) and Italian ryegrass (R2 = 0.16, Y = 3887 - 23.8*Yield), where Y is weed density (# emerged/m2) and Yield is kg/ha. These preliminary results are very promising and may have huge potential to improve weed management and combat herbicide resistance across enormous acreage. Future research over the following spring within this study will evaluate winter wheat and weed performance planted after soybean.

CROP RESPONSE TO SIMULATED DICAMBA DRIFT IN FIELD-GROWN SNAP BEANS. M Wasacz*, BL Carr, M VanGessel, DJ Mayonado, LM Sosnoskie, TE Besancon; Rutgers University, Chatsworth, NJ, University of Delaware, Georgetown, DE, Affiliation Not Specified, Hebron, MD, Cornell University, Geneva, NY

ABSTRACT

The spread of herbicide resistant Palmer amaranth (*Amaranthus palmeri* S. Watson) and horseweed (*Erigeron canadensis* L.) has contributed to increasing the acreage of dicamba-tolerant soybean in the mid-Atlantic region. Dicamba is a synthetic auxin herbicide that is prone to volatilization and drift, which may put neighboring dicamba-sensitive crops at risk. This research seeks to determine the sensitivity of field-grown snap beans (*Phaseolus vulgaris*) to simulated drift rates of dicamba. In Summer 2020, similar field research was conducted in Bridgeton, NJ and Geneva, NY. Snap beans of 'caprice' and 'huntington' variety were exposed to six micro-rates of dicamba (fractions of the soybean recommended label rate [560 g a.e. ha^{-1}]: 1/250, 1/500, 1/1000, 1/5000, 1/10000, and an untreated control) at two application timings (V3 and R1). After applications, visual injury ratings were taken 2 and 4 weeks after treatment. Snap beans were harvested, sorted, and weighed. Fresh plant biomass was measured in each plot. Snap bean plants showed injury symptoms including leaf deformation, mild chlorosis, stunting, epinasty, and bean deformation. Visual injury ratings suggest that plant injury was more severe at the V3 stage, and that all treatments except for the 1/10000 rate were significantly different than the untreated control plots. Visual injury at the V3 application timing was greater than 50% for the 1/250 and 1/500 treatments. Total bean yield data suggests that application at 1/250, 1/500, and 1/1000 of the labeled rate significantly reduced yields by 63%, 48%, and 26% on average, respectively, compared to the untreated control. However, there was no significant effect of application timing on bean yield. Biomass data indicate that plant biomass is not significantly affected by sublethal rates of dicamba, with the exception of the highest rate applied (1/250 of the labeled rate) which was significantly lower than the untreated control, with a 32% biomass reduction on average. These preliminary results suggest that snap bean is a highly sensitive crop to dicamba injury. Future studies involve replication in 2021 and potentially with other varieties of snap beans.

UNDERSTANDING STRESS ADAPTATION THROUGH DNA METHYLATION IN SORGHUM BICOLOR. G Sharma*, J Barney, J Westwood, S Askew, DC Haak; Virginia Tech, Blacksburg, VA, Affiliation Not Specified, Blacksburg, VA

ABSTRACT

Plants are sessile organisms and cannot respond rapidly to environmental fluctuations like animals or humans. Therefore, plants have developed effective survival strategies that are compatible with their sessile lifestyle, which range from short term physiological changes to long term genomic adaptations. One of the important mechanism of gene regulation in response to stresses is DNA methylation, addition or removal of a methyl group to cytosine nucleotides. DNA methylation is thought to contribute greatly to a plant's ability to respond to stress. There are several examples indicating that the level of DNA methylation increases or decreases in response to stress. Weed management practices such as herbicide and cultural methods acts as a stress to weeds. Moreover, overreliance on few herbicide modes of action leads to herbicide resistance in weeds. Understanding DNA methylation can aid our understanding of weed stress adaptation and development of herbicide resistance. Thus, we are working with *Sorghum bicolor*, closely related to Shattercane, a problematic weed in US. We administered sub-lethal doses of glyphosate, trifloxysulfuron, clipping, and shading to determine if methylome changes are shared or unique among stress responses. Methylation occurs in all cytosine sequence contexts of plant DNA: CG, CHG and CHH (H represents A, T or C). The tissues from control and stressed plants were collected at standardized maturation levels and subjected to whole-genome, bisulfite sequencing (WGBS). We found that highest levels of methylation in CG (20.5%) context followed by CHG (7.5%) and CHH (4.8%) context. Our results are in accordance to previous studies and indicate that methylation at promoter sites (CpG islands) are crucial for stress adaptations. DNA hypomethylation refers to the loss of the methyl group in the 5-methylcytosine nucleotide, whereas hypermethylation refers to the addition of the methyl group. The gain and loss of the methyl group can change gene expression. Glyphosate and shade stressed plants showed hypermethylation at transposable and repeat elements. Whereas hypomethylation in glyphosate stressed plants was recorded at promoter and genic regions. These results indicated that DNA methylation patterns are non-random and unique for each stress response. Further analyses will elucidate how location of methylation in genome aids in stress adaptation.

HOME CLIMATE AND LAND USE AFFECT SEED, BUT NOT RHIZOME, COLD TOLERANCE OF A PERENNIAL INVADER. VT Lakoba*, J Barney; Virginia Tech, Blacksburg, VA

ABSTRACT

Throughout the world's temperate regions, extreme cold plays a key role in range boundaries of terrestrial plants. Winter survival is critical to their persistence, but not all tissues or structures are equally susceptible to frost kill and, therefore, limiting to the species' distribution. There is also reason to expect intraspecific variation in cold tolerance both within and among tissue types. To address this problem, we experimentally found freezing tolerances of two overwintering propagule types -- seeds and rhizomes -- of Johnsongrass (*Sorghum halepense*), a globally distributed invasive species and agricultural weed. We compared tolerance thresholds of 18 genotypes sampled from agricultural and non-agricultural habitats spanning the climatic extremes they occupy in the United States. Single node rhizome fragments had an average LT90 of -5.1C with no significant variation based on home climate or ecotype. Non-dormant seeds frozen at -85C suffered an overall decline in germination to 10% from 25% at 22C. While population origin did not affect seeds' response to any treatment temperature, germination proportion and speed of non-agricultural populations was higher than agricultural populations from the coldest climates, with a reversed relationship among warmest origin seeds. Overall, populations from the cold/dry and wet/warm parts of Johnsongrass's range showed greater and faster germination. The vast differences in cold tolerance between propagule types, as well as evidence for adaptation to land use and climate of its seeds, suggest that this invader's continental spread is likely limited by winter survival of rhizomes, but not seeds. These findings afford us deeper insight into the species' dispersal dynamics, as well as where to focus future research on mechanisms of its range limitation.

SOIL CARBON AMENDMENTS AS A MICROBIALLY MEDIATED WEED MANAGEMENT TOOL. M Gannett*, A DiTommaso, J Kao-Kniffin; Cornell University, Ithaca, NY, Cornell University, Dryden, NY

ABSTRACT

Weed management tools often also interact with other aspects of a cropping system, which may lead to positive or negative synergies. A potential method of weed management is through soil carbon (C) amendments, which may also affect soil health. Soil carbon amendments stimulate soil microbial growth, which immobilizes nitrogen. Agricultural weeds, many of which grow well in high nitrogen soils, will have reduced growth in these high carbon, low available nitrogen soils. In a field experiment, we implemented five amendment treatments increasing in carbon: an unamended weed-free control, an unamended weedy control, rye straw adding about 4,000 kg C ha^{-1}, sawdust adding about 5,500 kg C ha^{-1}, and rye-straw and sawdust combined adding about 9,500 kg C ha^{-1}. We planted corn and soybean and each treatment was replicated 5 and 6 times respectively. The corn was band fertilized once in the spring, and the soybean fixed its own nitrogen. We measured aboveground weed biomass, aboveground crop biomass, and soil health using the Cornell Soil Health test. Both weed and crop biomass were significantly different based on amendment treatments. The crop to weed biomass per meter ratio was highest in the weed-free control plots for both corn and soybean. As more carbon was added to the soil, the crop to weed biomass ratio of corn decreased, but increased for soybean, possibly because soybean fixes its own nitrogen and was less negatively affected by the carbon. Soil health ratings improved as more carbon was added to the soil, indicating that for legumes such as soybean, soil carbon amendments may be able to help manage weeds and improve soil health without negatively affecting yield.

EVALUATION OF NON-CHEMICAL PRODUCTS AND ALTERNATIVE COMPOUNDS FOR ANNUAL BLUEGRASS CONTROL AND HOST SAFETY IN PENNSYLVANIA. K Tang*, J Kaminski; Penn State University, State College, PA, Penn State Univeristy, State College, PA

ABSTRACT

Annual bluegrass (*Poa annua* L., ABG) is one of the most prevalent turfgrass weeds in the United States. Most ABG management programs rely on herbicides but the species is resistant to at least 9 different herbicide sites of action. The objective of this research was to elucidate the efficacy of alternative control products on controlling ABG and to determine their safety when applied to creeping bentgrass (*Agrostis stolonifera* L., CBG). The study was initiated in 2020 at the Joseph Valentine Turfgrass Research Facility in University Park, PA on an ABG and CBG research fairway. Plots measured 3 x 6 ft and arranged as a randomized complete block design with 4 replications. Treatments included one control and 15 products including: Fiesta (27 % Iron HEDTA, 32.6 L ha^{-1}), AXXE (40 % Ammonium Nonanoate, 81.4 L ha^{-1}), Avenger (70 % d-limonene, 162.8 L ha^{-1}), Vineger30 (30 % Vinegar, 814.1 L ha^{-1}), Vineger40 (40 % Vinegar, 814.1 L ha^{-1}), Verdesian (Ferrous Sulfate Heptahydrate, 97.64 kg ha^{-1}), Alldown (23 % Acetic Acid, 14 % Citric Acid, 357.8 L ha^{-1}), FinalSan (22 % Ammoniated Soap of Fatty Acids, 162.8 L ha^{-1}), Scythe (57 % Pelargonic Acid, 3 %Fatty Acids, 81.4 L ha^{-1}), WeedZap (45 % Clove Oil, 45 % Cinnamon Oil, 40.7 L ha^{-1}), Suppress (47 % Caprylic Acid, 32 % Capric Acid, 24.4 L ha^{-1}), BurnOut (8 % Clove Oil, 24 % Citric Acid, 268.6 L ha^{-1}), ThymeGuard (23 % Thyme Oil, 4.1 L ha^{-1}), WeedSlayer (6 % Eugenol, 16.3 L ha^{-1}), and Phydura (10 % Clove Oil, 203.5 L ha^{-1}). Treatments were applied once on 24 June and the study was repeated on 14 Sept for each turfgrass species. Turfgrass injury was visually rated for 3 weeks on a 0 to 100% scale where 0 = no injury and 100 = entire plot area brown or dead. Vineger30 and Vineger40 injured ABG most when applied in the summer but caused significant injury to CBG at both timings. AXXE, Avenger, FinalSan, Alldown, Scythe, and BurnOut resulted in the greatest injury to ABG with the least amount of injury to CBG in the summer. Except for the Vinegar treatments, treatments applied in the autumn did not injury either turfgrass species when compared to the nontreated control. Future research should investigate various application timings and rates as well as the long-term control of ABG.

EVALUATING ELECTRICAL AND MECHANICAL METHODS FOR PALMER AMARANTH (AMARANTHUS PALMERI) CONTROL. LD Moore*, KM Jennings, D Monks, MD Boyette, DL Jordan, RG Leon; North Carolina State University, Raleigh, NC

ABSTRACT

Using electricity to selectively control weeds taller than the crop canopy has recently sparked interest in many conventional and organic growers as an alternative to hand weed removal. In addition, machinery to mechanically remove tall weeds from crops are available. However, little research has compared these weed control strategies or evaluated application timings that will result in optimal weed control and crop yield and quality. Palmer amaranth (*Amaranthus palmeri* S. Wats) is a problematic weed in the United States and can reach heights over 2 m tall, whereas both peanut and sweetpotato grow less than 0.5 m in height. Thus, field studies were conducted to determine recommendations for electrical and mechanical Palmer amaranth control in peanut and sweetpotato. Treatments were arranged in a three by four factorial of electrical (The Weed Zapper), mechanical (Bourquin Organic Weed Puller / Roguing Machine), or hand removal applied when Palmer amaranth were less than 0.3, 0.6, 0.9, or 1.2 m in height. In addition, a nontreated check was included for comparison. All treatments resulted in at least 78% Palmer amaranth control 8 wk after planting (WAP). Palmer amaranth control from electrical treatments were similar to the hand removal treatments when applied before weeds were 0.6 m in height. All electrical and mechanical treatments yielded less than early (Palmer amaranth less than 0.3 m in height) hand removal. Electrical and mechanical weed control can be effective alternatives to hand Palmer amaranth removal but should be incorporated into management systems rather than utilized as the primary weed management strategy.

CHARACTERIZATION OF HERBICIDE RESISTANT COMMON RAGWEED FROM THE MID-ATLANTIC REGION. FJ D'Amico*, TE Besancon, LS Shergill, M VanGessel; University of Delaware, Newark, DE, Rutgers University, Chatsworth, NJ, University of Delaware, Georgetown, DE

ABSTRACT

Common ragweed (*Ambrosia artemisiifolia L*) is a problematic weed throughout the United States and Canada and has been introduced in many areas around the world. Common ragweed can cause substantial crop losses and is a severe allergen to many people. In recent years control of common ragweed has been impeded by resistance to certain herbicides. The first confirmed case of resistance was to atrazine (group 5) in Canada (1976) associated with maize. Since then, resistant populations have been identified in numerous states, most associated with soybean cropping systems. In an effort to better characterize the extent of resistance in the Mid-Atlantic region, seeds were collected from multiple sites with suspected resistant populations and three areas with susceptible populations. The collection sites included eastern Virginia, multiple areas of Delmarva, southern Maryland, and New Jersey. Most collection sites were associated with soybean cropping systems which represent approximately 1.3 million acres in those states. In total 34 accessions were tested for postemergence resistance to glyphosate, cloransulam and fomesafen. All herbicides were applied at 1X, 2X and 4X local field rates. Of the 31 resistant accessions, none were controlled by glyphosate at 1X rate, only four of the accessions were controlled cloransulam at 1X rate. Fomesafen at 1X rate failed to control 12 of the accessions. Since no accession with suspected resistance was controlled by glyphosate, all 26 accessions resistant to cloransulam demonstrated two-way resistance. Likewise, all accessions resistant to fomesafen were also resistant to cloransulam, demonstrating three-way resistance. Select accessions were also tested for resistance to preemergence applications with group 2 (cloransulam and chlorimuron) or group 14 (fomesafen and sulfentrazone) herbicides. One of the accessions exhibited no effect at 2X rate of group 2 herbicides applied PRE, while two others demonstrated lower levels of resistance. Two accessions had multiple plants emerge and survive at 1X rate of group 14 herbicides. Two-way and three-way resistant populations of common ragweed may be more prevalent throughout the Mid-Atlantic region than previously recognized. The accessions studied exhibited resistance to both POST and PRE applications, complicating resistance management. Additional research will examine the genetic source of this resistance.

GERMINATION AND HERBICIDE SUSCEPTIBILITY OF OFFSPRING FROM PALMER AMARANTH (AMARANTHUS PALMERI) SURVIVING GLUFOSINATE APPLICATIONS. EA Jones*, DJ Contreras, M Fajardo Menjivar, DE Salazar, RG Leon, W Everman; North Carolina State University, Raleigh, NC

ABSTRACT

Experiments were conducted to determine the germination and susceptibility of offspring from Palmer amaranth (*Amaranthus palmeri*) plants surviving select glufosinate treatments. Seeds were collected from Palmer amaranth plants surviving non-treated, post, late post, and post followed by late post glufosinate applications at three different locations. Fifty seeds were counted from each Palmer amaranth accession and timing survived then placed into a petri dish containing 7.5 ml of de-ionized water plus blotter paper. The petri dishes were then placed into a germination chamber for 14 days with emerged seedlings counted and removed daily. Differential germination was observed across experimental runs and Palmer amaranth accessions. No observable patterns were detected between Palmer amaranth accession or timing survived. The differences were likely attributable to a confounding genetics*environment interaction. Seeds were sown in a greenhouse and treated with 1/8-2X of the rate of glufosinate (590 g ai ha^{-1}) applied in the field when the plants were approximately 7 cm in height. Visual control ratings were conducted weekly for three weeks. All plants treated with glufosinate were controlled within the three-week evaluation period. The results of the experiment provides evidence that the offspring of Palmer amaranth surviving select glufosinate applications do not exhibit differential germination or reduced susceptibility to glufosinate after one growing season.

PLANT HORMONES REDUCE BENTAZON INJURY TO SWEETPOTATOES, WITHOUT IMPACT ON HERBICIDE EFFICIENCY. GA Caputo*, MA Cutulle; Clemson University, Clemson, SC, Clemson University, Charleston, SC

ABSTRACT

Weed management is essential to maximize sweetpotato (Ipomoea batatas L. (Lam)) production. There are no selective herbicide registered in sweetpotato to suppress broadleaves and nutsedge in postemergence (POST). Expansion of bentazon and mesotrione herbicide labels would be beneficial to sweetpotato growers. Two experiments were conducted, the first experiment evaluated the response of sweetpotato cultivars 'Beauregard' and 'Covington' to bentazon and mesotrione when melatonin, 24-epibrassinolide, or ascorbic acid (AsA) were included in the tank-mix. The second experiment evaluated the efficiency of bentazon and mesotrione in control of yellow nutsedge (*Cyperus esculentus* L.) and palmer amaranth (*Amaranthus palmeri* S. Wats.) when different doses of melatonin, 24-epibrassinolide, or AsA were added into tank-mix with each herbicide. In the first experiment, when treated only with herbicide, 'Beauregard' exhibited lower injuries than 'Covington' to both herbicides used. Treatments using only AsA and plant hormones were not harmful to either sweetpotato cultivar. At the lowest dose of bentazon and mesotrione, the addition of compounds in tank-mix significantly reduced percent injury and increased plant tolerance. In the second experiment, the addition of plant hormones and AsA in tank-mix had a no-antagonistic effect on herbicide efficiency, exhibiting similar levels of injuries as herbicides application alone. These results suggest that the use of plant hormones and AsA could improve sweetpotato tolerance to POST applications of bentazon and mesotrione without reducing herbicide efficiency.

WEED MANAGEMENT PROGRAMS FOR REDUCED-TILLAGE SWEETPOTATO. SC Smith*[1], KM Jennings, D Monks, DL Jordan, MR Schwarz, C Reberg-Horton; North Carolina State University, Raleigh, NC, Affiliation Not Specified, Raleigh, NC

ABSTRACT

Sweet potato is an economically important commodity for the southeastern U.S. and California. Continuous tillage is a requirement in current production systems but is detrimental to soil structure and decreases soil organic matter content. The utilization of a high-residue rye cover crop mulch in sweetpotato production has the potential to reduce the need for and impact of tillage. Before reduced-tillage systems can be adopted, weed management programs need to be evaluated for efficacy. Thus, field studies were conducted at the Horticultural Crops Research Station in Clinton, NC and Caswell Research Station in Kinston, NC in 2019 and 2020, respectively. Treatments included two production systems (conventional, or reduced-tillage with rye cover crop). Herbicide program treatments included flumioxazin, fomesafen, indaziflam, or bicyclopyrone pre-transplant, with or without clomazone after transplanting, and S-metolachlor with or without linuron at 2 WAP. Herbicide programs including linuron at 2 WAP had the greatest broadleaf weed control in the reduced-tillage system. All programs resulted in similar broadleaf weed control in the conventional tillage system. Flumioxazin followed by clomazone followed by S-metolachlor tank mixed with linuron resulted in similar total storage root yield as the weed-free check in the reduced-tillage system. All herbicide programs in the conventional tillage system resulted in similar total storage root yield as the weed-free check.

INTEGRATION OF HALOSULFURON AND ANAEROBIC SOIL DISINFESTATION FOR WEED CONTROL IN TOMATO. G Singh*, MA Cutulle, HT Campbell, MA Cutulle; Clemson University, Clemson, SC, Clemson University, Charleston, SC

ABSTRACT

Weeds are a major yield-limiting factor in specialty crop production. Herbicides have been an important tool for weed control after the soil fumigant Methyl bromide (MeBr) was removed from the market due to health and environmental concerns. However, limited access to effective herbicides in specialty crops, and rising herbicide resistance cases hint at an uncertain future for herbicide-based weed management. Additional strategies are needed that can effectively shift present weed control tactics towards an integrated system to reduce over-reliance on chemical herbicides. Anaerobic soil disinfestation (ASD) has the potential to fit into current weed management practices. ASD involves the application of a carbon source, irrigation to field capacity, and covering the soil with a plastic tarp. Changes in the soil microbial communities and production of volatile organic compounds during anaerobic decomposition have been reported as major control mechanisms of pathogens and nematodes. ASD was initially developed to control soilborne pathogens but has also been reported effective against several weeds. Two greenhouse studies followed by a field experiment performed at joint Clemson University and USDA vegetable research center, Charleston, South Carolina with three replicates to evaluate two carbon sources (mustard meal + molasses (MMM) or no carbon amendment), three herbicide treatments (halosulfuron applied pre-ASD, halosulfuron applied post-ASD, and no herbicide), as well as two plasticulture treatments (sealed or not sealed). In greenhouse studies, each experimental unit was seeded with *Cyperus esculentus* (Yellow nutsedge) tubers and *Amaranthus palmeri* (Palmer amaranth) seeds. For non- sealed treatments buckets were left completely uncovered in greenhouse, while in field study plots were covered but aerated on the side of the bed. In both greenhouse and field studies, treatment rates were same. Soil treatments included 13.5 m^3 ha^{-1} of molasses and 2.137 Mg ha^{-1} of mustard meal. Halosulfuron with 75% active ingredient was used at a rate of 70.02 grams/hectare. ASD was performed for 4 weeks, and then terminated by removing polyfilm sealing in greenhouse, while poked holes in field study. In both studies, MMM and polyfilm sealing with either halosulfuron application resulted in 98% or greater control of *Cyperus esculentus*(Yellow nutsedge). Pre-halosulfuron applications, along with carbon source amendments, improved *Cyperus esculentus* (Yellow nutsedge) control and had no negative impact on the development of anaerobicity in ASD-treated plots. Up to 100% control of *Amaranthus palmeri* (Palmer amaranth) and *Echinochloa crussgalli* (Barnyardgrass) was exhibited in MMM + Polyfilm sealing alone or with either of halosulfuron application timings. In some ASD plots allelopathetic effects in terms of plant injury and less plant vigor were observed in Tomato plants. Overall, MMM + halosulfuron Post-ASD treatment reported most effective in suppressing weeds and better yield in Tomato production.

CHARACTERIZING FLUMIOXAZIN AND IMAZAPIC RESIDUAL ACTIVITY AND SOIL FATE USING FIELD AND GREENHOUSE BIOASSAYS WITH B. CARINATA. RG Leon, ME Camacho*; North Carolina State University, Raleigh, NC

ABSTRACT

Carinata (*Brassica carinata* A. Braun) has become a promising alternative crop for biofuel production in the Southeastern USA. However, there are concerns of carryover issues caused by soil persistent herbicides, which are used in rotational crops and to which carinata is susceptible. The present study evaluated the potential carryover risk for carinata of imazapic and flumioxazin. Field trials were established in two different sites (sandy and sandy loam soils) in North Carolina, USA. Applications were done to bare ground with the recommended label rate of imazapic 70 g ai ha^{-1}) and flumioxazin (107 g ai ha^{-1}) at 24, 18, 12, 6 and 3 months before carinata planting. In addition, those herbicides were applied in preemergence (PRE) right after carinata planting at 1X, 0.5X, 0.25X, 0.125X and 0.068X of the label rate. A non-treated control was included for comparison. Intact soil cores were taken from the field plots, aiming to establish a greenhouse bioassay to assess herbicide movement in the soil and its effect on carinata plants. No differences in crop establishment were observed for applications of those herbicides done longer than 3 months before planting. Imazapic applied PRE presented differences while comparing with control, with reduction up to 40% in plant survival. Flumioxazin applied PRE reduced carinata establishment causing the lowest plant survival (12.3% to 0.3%). Both herbicides applied in the 0.068X caused considerable plant mortality. The bioassay indicated that imazapic moved deeper in the soil profile allowing plants to get established and grow for some time before they reach the herbicide layer and exhibit injury symptoms. This vertical movement was more pronounced in the sandy soil. Conversely, flumioxazin remained at toxic concentrations in the top 3 cm of the soil.

OVERLAPPING RESIDUAL HERBICIDES FOR WEED CONTROL IN NO-TILL PUMPKIN. KM Vollmer*, K Nichols, D Lingenfelter, JM Wallace, M VanGessel, BA Scott; University of Maryland, Queenstown, MD, University of Maryland, Derwood, MD, Penn State University, University Park, PA, University of Delaware, Georgetown, DE

ABSTRACT

Weed control in pumpkin (*Cucurbita pepo* L.) can be challenging due to several factors. These include the wide row spacing needed for pumpkin production, a longer growing season, the prevalence of no-till in pumpkin production, and a lack of effective herbicides. As a result, new weed control strategies must be evaluated. Overlapping herbicides is a technique that involves sequential applications of soil-applied residual herbicides in order to overlap the activity of two or more herbicides before the activity of one herbicide dissipates. A study evaluated weed control efficacy and pumpkin response to *S*-metolachlor applied POST at two different rates and application timings. The study was repeated at four locations in Delaware, Maryland, and Pennsylvania over two years. Herbicide treatments included, ethalfluralin (1.3 kg ha^{-1}) applied PRE only, or ethalfluralin followed by *S*-metolachlor (0.8 kg ha^{-1} or 1.6 kg ha^{-}1) applied 2 wk or 4 wk after planting. Treatments of ethalfluralin + fomesafen (0.4 kg ha^{-1}) applied PRE, a weed free check, and an untreated check were included for comparison. No injury was observed following POST *S*-metolachlor applications, however minor stand loss was observed with ethalfluralin + fomesafen. No herbicide treatment controlled pigweed (*Amaranthus*) species and large crabgrass (*Digitaria sanguinalis* [L.] Scop.) greater than 90% late in the growing season. Ethalfluralin followed by 1.6 kg ha^{-1} *S*-metolachlor applied 2 wk after planting and ethalfluralin + fomesafen PRE controlled pigweed species greater than all other herbicide treatments. All ethalfluralin followed by *S*-metolachlor treatments controlled large crabgrass better than Ethalfluralin PRE regardless of rate or application timing. Yields were greater when 1.6 kg ha^{-1} *S*-metolachlor was applied 2 WAP compared to all other herbicide treatments, and were comparable to the weed free check. Our results show that *S*-metolachlor does not cause adverse crop injury when applied POST as an overlapping residual treatment in pumpkin. While this tactic improved late season weed control, it was best utilized when residual herbicides were applied closer to planting. Growers should be aware that *S*-metolachlor controls a limited weed spectrum and will not control emerged weeds. Therefore, overlapping residual herbicides should not be utilized as a stand-alone tactic, and integrated tactics, such as weed scouting, need to be incorporated for optimal weed control.

WEED FREE DURATION AND VARIETAL IMPACT ON HEMP (CANNABIS SATIVA L.) BUD YIELD IN COASTAL SOUTH CAROLINA. HT Campbell*, MA Cutulle, B Ward; Clemson University, Charleston, SC

ABSTRACT

With 114 licensed hemp (*Cannabis sativa L.*) growers in South Carolina, hemp is becoming more agriculturally significant across the state. Little research has been conducted exploring the impact of weed competition on hemp grown in South Carolina. Field experiments were conducted in 2019 and 2020 to assess dry bud yields of 6 hemp varieties commonly grown in South Carolina (Cherry Wine, Hurricane, Cherry Blossom, Therapy, T1, and Boax) when kept weed-free for various periods of time (0, 1, 2, 4, and 6 weeks). The trials were configured as a strip plot design with three replications. Weeding was achieved by manually hoeing around each plant approximately 3 feet from the base of the stem. The row middles, consisting mostly of grasses, were mowed once per week. Dry bud weights for each variety were significantly greater after 1 week weed-free compared to the non-weeded plots across all varieties in 2019, with the exception of Therapy. Among all varieties, Therapy had the lowest dry bud weights across both trial years for each weed-free period, minus the 2 week weed-free treatment in 2020. No significant differences in dry bud weights were observed for any of the 6 varieties between 1 week weed-free and 6 weeks weed-free in 2019, while four of the varieties achieved peak dry bud yields between 2-4 weeks weed-free (Boax, Cherry Blossom, Cherry Wine, and T1). The trial year 2020 saw significant increases in dry bud yields across all varieties, minus Therapy, between 1 week weed-free and 4 weeks weed-free, while there were no significant differences observed between 4 and 6 weeks weed-free for any of the 6 varieties. Based on this study, growers should aim to keep hemp production fields weed-free 1 week, at minimum, to maximize bud yields. For most hemp varieties, growers should aim to maintain weed-free settings for 2-4 weeks if possible, while extending total weed control beyond 4 weeks may not increase bud yields an acceptable amount to justify increased input costs.

OVERLAPPING NAPROPAMIDE AND MESOTRIONE APPLICATIONS IN CRANBERRY FOR SUPPRESSING CAROLINA REDROOT. TE Besancon, BL Carr*, M Wasacz; Rutgers University, Chatsworth, NJ

ABSTRACT

The perennial nature of cranberry (*Vaccinium macrocarpon* Aiton) production predisposes the crop to a diversity of weeds ranging from herbaceous to woody perennial species. Carolina redroot [*Lachnanthes caroliniana* (Lam.) Dandy] is a troublesome perennial weed of New jersey cranberry bogs that severely impacts fruit yield and harvest efficiency. Field studies were conducted in 2020 in Chatsworth, NJ, to evaluate Carolina redroot control and cranberry tolerance response to herbicide programs based on overlapping of PRE followed by POST applications. Treatments included napropamide at 6.7 or 10.1 kg a.i. ha^{-1} single- or split-applied PRE alone, or followed by single or double POST mesotrione at 0.3 kg a.i. ha^{-1}. Dichlobenil PRE at 4.4 kg a.i. ha^{-1} and equally split was also included. PRE applications were made immediately after draining of winter flooding water (mid-April), and eventually renewed 30 days later. POST treatments were first applied when Carolina redroot reached cranberry canopy height, and eventually renewed 30 days later. Napropamide applied at 6.7 kg a.i. ha^{-1} with POST applications of mesotrione provided 44% Carolina redroot control 120 DA-A. Splitting this rate into two unequal applications (4.5 fb 2.2 kg a.i. ha^{-1}) decreased Carolina redroot density nearly 63% compared to the weedy check. Napropamide applied at 10.1 kg a.i. ha^{-1} and followed with mesotrione POST averaged 47% Carolina redroot control 120 DA-A, and specifically reduced 76% the number of plants 76% when two mesotrione applications were included. Number of flower stalks produced was 93% reduced when herbicides were applied, regardless of treatment. No flower stalks emerged when napropamide applications where followed by one or two mesotrione applications. As a result of crop injury (chlorosis), plots in which dichlobenil was applied yielded 47% less than those treated with napropamide. An average 30% yield decreases was noted with split-applied napropamide at 10.1 kg a.i. ha^{-1} compared to a single application in early spring, regardless of POST mesotrione applications. No yield decrease occurred with split application of napropamide at 6.7 kg a.i. ha^{-1}.

VARIETAL TOLERANCE AND WEED CONTROL WITH SULFENTRAZONE IN SWEET BASIL.

BL Carr, M Wasacz, TE Besancon*; Rutgers University, Chatsworth, NJ

ABSTRACT

Sweet basil (*Ocimum basilicum* L.) is a culinary herb that attracts renewed interest in the Mid-Atlantic region with the commercial release of four Rutgers Downy Mildew Resistant cultivars in 2020. However, basil weed control remains challenging because of crop sensitivity to early weed competition and the limited number of effective and safe herbicide options. Field studies were conducted to evaluate sweet basil tolerance to various PRE herbicides in 2019, and to determine a strategy combining crop safety and weed control efficacy in 2020. Treatments in 2019 included sulfentrazone at 150 g ha^{-1}, linuron at 560 g ha^{-1}, ethalfluralin at 840 g ha^{-1}, fomesafen at 150 g ha^{-1}, bensulide at 5.6 kg ha^{-1}, pronamide at 1.1 kg ha^{-1}, and diuron at 1.1 kg ha^{-1} compared to labelled napropamide at 2.2 kg ha^{-1}. In 2020, sulfentrazone alone (150 g ha^{-1}) alone or mixed (110 or 150 g ha^{-1}) with napropamide at 1.7 kg ha^{-1} was applied at seeding. Treatments also included sulfentrazone (150 g ha^{-1}) delayed 5 d after seeding and napropamide (2.2 kg ha^{-1}) at seeding. 2019 results indicated improved broadleaf weed control with sulfentrazone, diuron, linuron and fomesafen compared to the napropamide standard. Severe inhibition of basil seed germination or crop stunting observed with diuron, fomesafen, and bensulide caused 75% to 100% fresh yield reduction compared to a hand weeded check. By contrast, sulfentrazone or linuron did not impact basil fresh yield. In 2020, hairy galinsoga and stinkgrass control 55 DAT remained = 85% regardless of sulfentrazone treatment but did not exceed 50% with napropamide alone. Delaying sulfentrazone application after the onset of basil seed germination significantly reduced the number of emerged seedling compared to sulfentrazone applied at seeding. Fresh yield was reduced 46% with delayed sulfentrazone application compared to a hand-weeded check. Fresh yield with sulfentrazone 110 g ha^{-1} plus napropamide 1.7 kg ha^{-1} at seeding yielded similarly to the hand-weeded check and increased fresh yield by 26% compared to napropamide alone at seeding.

COUNCIL FOR AGRICULTURAL SCIENCE AND TECHNOLOGY (CAST): ITS ROLE AND IMPORTANCE TO NEWSS AND NEPPSC. HA Sandler*; UMass Cranberry Station, East Wareham, MA

ABSTRACT

What is it? CAST, founded in 1972, defined its mission to assemble, interpret, and communicate credible, balanced, science-based information to policymakers, the media, the private sector, and the public. Its headquarters are located in Ames, Iowa and the nonprofit 501(c)(3) has approximately eight staff members. It has a Board of Directors (12 members: 5 elected officers and 7 liaisons), Board of Trustees (9 members from universities, industry, and government) and the Board of Representatives. The last board is where the NEWSS Representative to CAST sits along with approximately 40 other people from universities, industry, nonprofit and trade organizations and the BOD. All members of the BOR are members of one working group: Animal, Plant or Food Science and Safety. Working groups (WG) meet via conference call or teleconferencing monthly; all members attend the annual meeting held in the fall. WGs generate ideas and proposals that eventually become issue papers, commentaries or special publications. From inception to approved proposal usually takes a few months to 1 year. The publication process can take a few months to 1 year or more. What are its products? When a publication is ready for release, a 1-page, double-sided "Ag QuickCAST" is generated. Prior to Covid, the rollout of a new publication was a face-to-face event in Washington, DC. In 2020, CAST introduced its new papers via a free webinar; these have been very successful. Each week, it releases Friday Notes, an electronic publication featuring stories and highlights about all aspects of agriculture. CAST generates reports on all its activities; these documents are available on its web site: www.cast-science.org. Publications are available for free download from its web site. How can you help? NEWSS and NEPPSC members can support CAST by becoming a member; all subscriptions are tax-deductible due to CAST's nonprofit status. Many types of memberships are available including: Individual memberships at $100 per calendar year; recent graduates can become a member for $35 and students can become members at no cost. NEWSS and NEPPSC members can support CAST by generating ideas for papers and sharing those with any active representative of CAST. If asked to be an author or a reviewer, please consider being of service to this organization. Lastly, many people are unaware of the tremendous resource that is CAST. If you find a paper that is of interest to you, please share it with your colleagues and encourage them to visit the CAST web site www.cast-science.org.

WEED MANAGEMENT ON NEW CRANBERRY PLANTINGS. K Ghantous*, HA Sandler; UMass Amherst, UMass Cranberry Station, East Wareham, MA, UMass Cranberry Station, East Wareham, MA

ABSTRACT

Cultivated cranberry (*Vaccinium macrocarpon*) is a long-lived woody perennial. Plantings can lose productivity over time due to issues such as increasing weed and pest populations, loss of genetic integrity as off-type cranberry vines spread, and abiotic factors such as issues with grade and drainage. Growers strive to replant every 20 to 25 years. The first three to five years after planting are considered the establishment stage, and full productivity is not reached until the end of this stage. During establishment weed management is critical. Beds are planted with either unrooted cranberry vine cuttings that are pressed into the soil surface with a disc, or rooted plugs made from cuttings propagated in a greenhouse for several months. Over time, cranberry then spreads vegetatively to fill in the bare areas. Runners grow along the soil surface and put down roots. Regardless of method, large areas of bare soil are present on renovated beds for a long period of time before the cranberry vines can spread, root, and create a canopy. Weeds hinder cranberry colonization by creating a physical barrier to runners rooting, and competition for nutrients, space, light, etc. Poor weed management during initial years has long-lasting impacts. Available herbicide options considered safe on establishing cranberry plants include napropamide, mesotrione, and dichlobenil, however new tools for use during establishment are needed. Pronamide may be registered for use in cranberry in the near future, and safety for use on newly planted cranberries was unknown. A study was conducted in 2019 on a commercial cranberry farm located in South Carver, MA on two recently renovated beds. One bed was planted with disced-in vines (var. Mullica Queen) approximately 1 yr before pronamide treatments, and the other bed was planted with rooted cuttings (var. Haines) approximately 2 wk prior to treatments. Sets of 1-m^2 plots were established on each bed. Treatments were applied by CO_2-powered backpack sprayer. Herbicides were delivered in the equivalent of 3,740 L water ha^{-1} to simulate application by chemigation, the typical application method for pesticides in MA. Each plot received a single treatment (5.8 L ha^{-1} once, 2.9 L ha^{-1} once, 2.9 L ha^{-1} twice, or untreated) and all treatments were replicated four times. Plots were evaluated 2.5 mo after treatment (MAT). For the 1-yr old planting, there was no effect of treatment on total aboveground biomass or the aboveground biomass of current season growth when compared to the untreated control. For the newly planted plugs, there was no effect of treatment on cranberry vine cover when compared to the untreated control. All pronamide treatment significantly reduced weed cover compared to the untreated control on the 2-wk planting (no weeds present at time of treatment) and 5.8 L ha^{-1} once and 2.9 L ha^{-1} twice significantly reduced weed cover on the 1-yr planting (weeds present at time of treatment). A greenhouse study was conducted with unrooted cuttings, using the same pronamide treatments as the 2019 field study. 10-cm cuttings of 'Stevens' were planted in sand 3 wk prior to treatment (4 cuttings per treatment per rep). Treatments were replicated four times, and the study was conducted twice. Plants were collected 3 MAT and dried to measure above- and belowground biomass. There were no significant difference between treated and untreated plants for aboveground, belowground, or total biomass. Preliminary findings indicate that pronamide is safe for use on new cranberry plantings, and would provide growers with a new and effective weed management tool during the establishment of new beds.

75 YEARS OF THE NORTHEASTERN WEED SCIENCE SOCIETY. D Kunkel*, AE Gover[2]; AMVAC, Plainsboro, NJ, Penn State, University Park, PA

ABSTRACT

2021 marks the 75th Meeting of the NEWSS. In early 1947, the Director of the Cornell Experiment Station issued and open invitation to all interested parties to participate in a workshop on weed problems and to explore the feasibility of forming an organization of weed workers in the northeast. They convened on the Cornell campus on February 18-19, 1947. Robert D. Sweet of Cornell was the organizer and acting chairperson of the meeting. In attendance, there were 47 participants from colleges, experiment stations, and federal units, while 37 were from industry representing 22 companies. State experiment stations represented included: Main, New Hampshire, New York, Massachusetts, Rhode Island, Michigan, Ohio, Pennsylvania, New Jersey and Virginia. The Northeastern Weed Control Conference formally formed on February 19, 1947. The name changed to Northeastern Weed Science Society in 1970. The second meeting was held in New York City, January 12-13, 1948 where meetings continued each year until 1973, when the meeting began to rotate to different Northeastern cities. The membership reached a high of over 750 members in the mid-1960s. The conference supported the weed field days that started in the mid 1950's and the first student weed contest (Weed meet) was held in August of 1983 at the Wye Institute on Maryland's Eastern Shore. Today the society continues with over 100 members. The annual meeting has evolved into the the NorthEastern Plant Pest and Soils Conference, hosted by the NEWSS and attended by members of the NE societies of the American Phytopathological Society, American Society of Agronomy, Crop Science Society, Soil Science Society of America and the American Society of Horticultural Science. The importance of protecting US crops from weeds and other pest, that rob growers from crop yields and waste food, remains as important today as it was in 1947! *Extracted, in part from: Northeastern Weed Science Society History 1947-1995, Robert D. Sweet.*

UPDATE ON IR-4 WEED SCIENCE PROJECTS - FOOD CROPS. RB Batts*, JJ Baron, D Kunkel, MP Braverman; IR-4 Project HQ, NC State University, Raleigh, NC, AMVAC, Plainsboro, NJ, IR-4 Project Headquarters, Princeton, NJ

ABSTRACT

Update on IR-4 Weed Science Projects – Food Crops. Roger B. Batts[1], Jerry Baron[1], Michael J. Braverman[2], Daniel Kunkel[3]; [1]IR-4 Project, NC State University, Raleigh, NC, [2]IR-4 Project, Rutgers University, Princeton, NJ, [3]AMVAC Corporation, Plainsboro, NJ Data submitted by IR-4 led to just under 600 new uses in 2020, which is substantially fewer than in most years. Nineteen of these were for herbicides and plant growth regulators (including prohexadione calcium, sethoxydim, s-metolachlor, isoxaben, saflufenacil and 2,4-D) in a wide range of crops (alfalfa, globe artichoke, caneberry subgroup, hops, dill, rosemary, fig, chia, basil, sesame and intermediate wheatgrass). In 2020, IR-4 data petitions were submitted to EPA for ethalfluralin/stevia, glufosinate/fruiting vegetables, curcurbits, fig, avocado and hops, 2,4-D/clover and floransulam/grasses grown for seed. Sixteen new herbicide magnitude-of-residue studies began in 2020. The number of on-going herbicide Product Performance studies in 2020 was twenty. Development of Product Performance data (efficacy and crop safety research) to support labeling of new uses for specialty crop pest management tools continues to be an important priority in the IR-4 Project's annual research plan as the data are often required by registrants and states (e.g.California) to complete the registration process. The 2021 field research plan for herbicides and plant growth regulators includes fifteen residue studies and twenty-two continuing or new Product Performance studies. Five new weed control Integrated Solutions studies will also begin in 2021, including grapes, apples, blueberries, processing tomatoes, sweetpotatoes, haskap, and date palm. The Integrated Solutions Program is designed to address pest problems without solutions, resistance management, products for organic production and pesticide residue mitigation. The transition of IR-4 Headquarters to NC State University continues, but was slowed for various reasons during 2020. This relocation should be completed by September 2021.

Author Index

Northeastern Weed Science Society (NEWSS)
P.O. Box 25
Woodstown, NJ 08098
USA

ISBN 978-1-7138-4212-5

CURRAN ASSOCIATES INC.
proceedings.com